AF535073

Handbook
on
Oil and Gas Accounting

Handbook
on
Oil and Gas Accounting
by
Robert J. Koester

The Institute for Energy Development
Oklahoma City, Oklahoma

Handbook on Oil and Gas Accounting

First Printing: May, 1982
Library of Congress Catalog Card Number:
ISBN Number: 0-89419-196-9

Printed at IED Press, Inc., Oklahoma City, Oklahoma
UNITED STATES OF AMERICA

Editing by:
Peggy J. Durham, President,
The WORD PLACE Company,
Oklahoma City, Oklahoma

Illustrations by:
Joyce Walton,
J Walton Commercial Art
Oklahoma City, Oklahoma

DEDICATION

to Alice,
the spice of my life.

ABOUT THE AUTHOR

ROBERT J. KOESTER

Robert J. Koester, Ph.D., C.P.A., is Professor of Accounting at Texas Tech University where he directs the nation's only program leading to a Master's Degree in Oil and Gas Accounting. He also serves as Director of Texas Tech's Center for Oil and Gas Accounting and Management, and Continuing Education. His industry experience includes extended consulting assignments with ARCO, Exxon Company U.S.A., and the State of Alaska, as well as assignments with numerous other oil companies and CPA firms. Dr. Koester has published numerous technical articles on oil and gas accounting.

Table of Contents

Illustrations

Preface

This Handbook deals with contemporary financial accounting issues for oil and gas producing companies. Supplemented by problems and exercises, it can be used as a single volume textbook in a petroleum accounting course. It will also serve as a basic training aid for anyone needing exposure to the area. A detailed accounting background is not necessary to understand the material in this book.

I appreciate the assistance and encouragement from J. Dwain Schmidt, Lane Anderson, and Peggy Durham.

This volume is designed to be augmented by the Institute for Energy Development *Dictionary for Oil and Gas Accounting* and *Case Study Book for Oil and Gas Accounting*.

Robert J. Koester
March, 1982

Chapter 1

Purposes of Accounting and Financial Statements

Welcome to the exciting and challenging world of oil and gas accounting! The end product of oil and gas accounting is financial statements of oil and gas producing companies. The readers of this book undoubtedly will have various degrees of accounting background. Since this Handbook is designed to serve a multitude of purposes, the first chapter is devoted to a summary review of those accounting concepts and principles necessary to understand oil and gas accounting. This book assumes only that the reader has a basic understanding of the debit and credit mechanical process. All other accounting issues needed to understand the concepts in this book are discussed here. The experienced accountant will, of course, skip Chapter One and go to a chapter of specific interest.

FINANCIAL STATEMENTS

The term *financial statements* means different things to various people. It may refer to the audited annual report to stockholders, to an internal budget report or to a tax return. In this first section of the chapter we will identify the various financial statements which are typically prepared by an oil and gas producing company, and discuss their purpose and constraints. Please refer to Figure 1-1 which diagrams the different forms of statements we'll discuss here.

INTERNAL STATEMENTS

The primary internal financial statement in any organization is the *budget*. The budget is prepared as an anticipation of revenues and expenditures for a future period. The budget also normally includes a comparison of actual revenues and expenditures after the fact. It frequently takes the same form as other financial statements, such as an *income statement* and a *balance sheet*. Although internal statements may be in the same format as other financial statements, they may have some sections with more detail.

The primary characteristic of internal financial statements is that they are not subject to the constraints of any external authority, such as legal requirements on an external auditor's report.

The purpose of internal financial statements is to provide information for internal decision making. They represent the primary financial information tool of management.

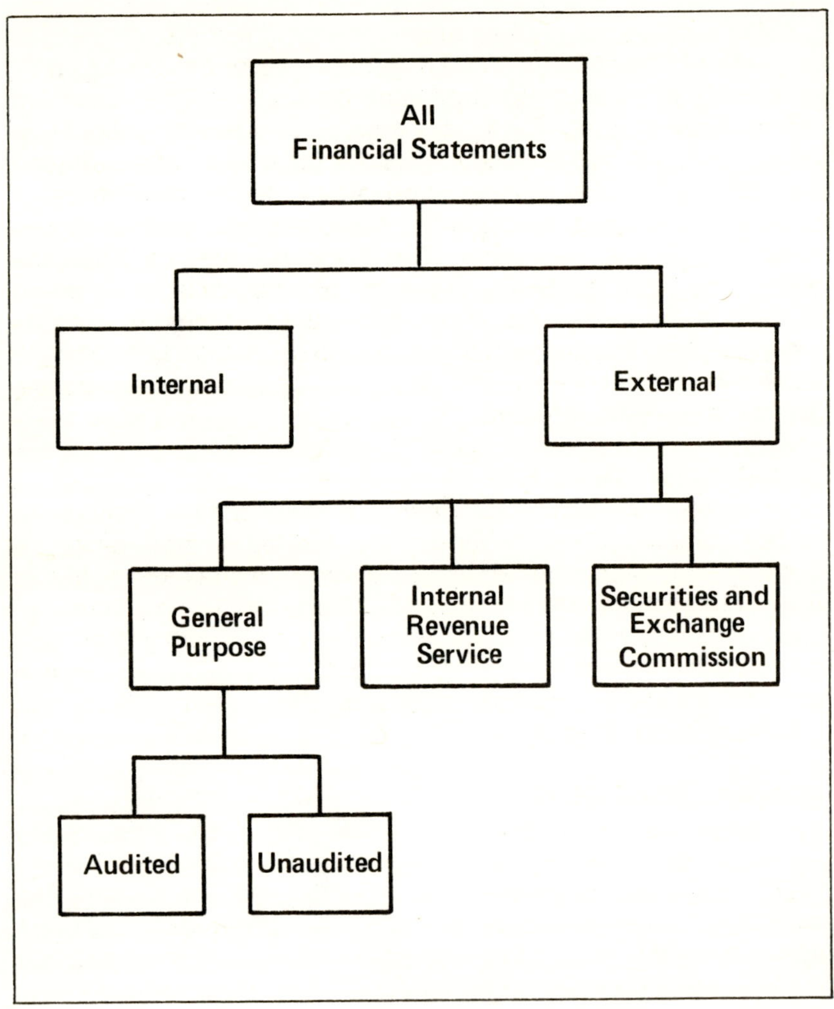

Figure 1—1

EXTERNAL FINANCIAL STATEMENTS

Any statement which contains financial information and is sent to parties external to the organization can be distinguished from internal financial statements. All external statements are prepared to meet the specific needs of those external users. They are divided into three general categories: general purpose financial statements,reports to the *Internal Revenue Service* (IRS), and reports to the *Securities and Exchange Commission* (SEC). Other reports, such as those to the *Department of Energy*, are prepared but they will not be covered here.

INTERNAL REVENUE SERVICE REPORTS

Most oil and gas producers must pay federal income taxes. Under the tax system in the United States, the taxpayer must report his own earnings and compute his taxable income under the applicable rules. The rules to determine this taxable income stem from several sources: the Internal Revenue Code, rulings by the U.S. Treasury Department, and various court decisions. These tax rules are very complex, highly complicated and substantially different from other accounting rules.

A taxpayer who has no other reporting requirements will frequently keep his accounting records in accordance with these rules, called "keeping his accounts on the *tax basis*". If such a taxpayer is then required to file another form of external report, he must convert the data to the rules that apply to that other report. The specific tax rules for oil and gas producing companies are outlined in Chapter Nine. The majority of this Handbook, however, deals with the other forms of external reporting requirements.

SECURITIES AND EXCHANGE COMMISSION

Companies whose stock is traded on one of the stock exchanges in the United States or one of the over-the-counter markets, and companies who offer for public sale large amounts of common stocks, bonds, or investments such as drilling funds are subject to reporting requirements of the Securities and Exchange Commission. This federal commission, appointed by the President, serves to make available to the investing public, information it feels is necessary to protect the interest of the investor.

Companies attempting to raise capital by new issues of common stock, bonds, or other forms of investment are required to file a *prospectus* with the S.E.C. This prospectus must be prepared in accordance with very carefully detailed rules, and usually is accompanied by a complete set of financial statements. Those companies whose stocks or bonds are traded on the open market must also file with the S.E.C. an annual report, called a *Form 10-K*, and a quarterly report, called a *Form 10-Q*. These reports are made available to the public and must be prepared within very carefully defined rules.

The basic vehicle the S.E.C. uses to implement a change in these rules for a Form 10-K or 10-Q is called an *Accounting Series Release* (ASR). There will be many references later on in this Handbook to such releases. The two primary ASR's (275 and 258) for oil and gas producing companies are reproduced in Appendix A.

We should note here that many companies are not subject to rules by the S.E.C. The corporation whose stock is closely held, that is, it is not publicly traded or sold, is usually not subject to the reporting requirements of the S.E.C. In the accounting literature the term *public company* refers to those companies that are traded on the open market and who are subject to the S.E.C. The term *private company* refers those proprietorships, partnerships, and closely held corporations, that are not subject to S.E.C. reporting.

GENERAL PURPOSE STATEMENTS

These statements include the annual report to stockholders, which is the most common form of general statement. Also included in this rather broad category are all other sets of financial statements prepared for external users. For example, banks frequently require a set of financial statements before considering a loan application.

General purpose financial statements consist of an income statement, a balance sheet, a statement of changes of financial position, and numerous other items of disclosure. Two complete sets of such financial statements are illustrated in Appendix B.

These statements may be audited or unaudited. Statements which bear an *independent auditor's opinion* have been reviewed by an independent certified public account who then expresses an opinion on those statements.[1] The opinion tells the statement reader whether or not the auditor feels the statements fairly pre-

[1]See pages 199 and 220 of Appendix B.

sent the financial position and the results of operations of the company. In addition, the auditor expresses an opinion as to whether or not the company has consistently followed generally accepted accounting principles in the preparation of those statements.

GENERALLY ACCEPTED ACCOUNTING PRINCIPLES

Generally accepted accounting principles refer to all the official pronouncements that have been processed or promulgated by the various accounting bodies, plus a rather loosely defined set of rules which are generally followed in the profession. The current private rule making body for the accounting profession is the *Financial Accounting Standards Board* (FASB). The board has recently issued several documents which specifically relate to oil and gas producing companies. These documents, called *Statements of Financial Accounting Standards,* are discussed in detail in subsequent chapters. A company which presents an audited set of financial statements, must follow all official Statements of Financial Accounting Standards which are in force on the date the financial statements are released. This is true for both private and public companies.

Many times, financial statements will not be *audited*. They may be marked *unaudited* on the face of the statements and there will not be an auditor's opinion accompanying those statements. If a certified public accountant has prepared the statements but has not performed certain auditing procedures he must include the phrase "unaudited" on the face of the statements and must add his own statement that he has not, in fact, audited the statements. If the statements have been prepared by someone other than a certified public accountant, no such declaimer of opinion is necessary. Banks will sometimes accept financial statements which have not been audited, because the cost of an audit can be prohibitive.

If financial statements are unaudited, the reader must be alert to two dangers. First, there has been no independent verification that the amounts have been properly recorded. Second, there is no assurance to the financial statement reader that the statements have been prepared in accordance with generally accepted accounting principles.

In this Handbook, we will usually refer to general purpose financial statements which have been audited, such as the annual report to stockholders. In those cases where the discussion centers on others types of statements, the specific requirements will be explained.

THE BASIC FINANCIAL STATEMENTS

The income statement and balance sheet provide the heart of the financial statement disclosure package. As the reader can easily tell from Appendix B many other items of information are included in a complete set of financial statements. However, in this discussion, we will concentrate on these two primary statements.

THE INCOME STATEMENT

Refer to Figure 1-2 for an illustration of an income statement. Most income statements contain more information than this, but the essential features of this statement are highlighted here. The purpose of an income statement is to describe the results of an entity's operations over a specific period of time. The time period is generally a year, although income statements are also frequently prepared for a quarter of a year, or on a monthly basis. An income statement can be prepared for any period of time.

Revenue refers to the cash or the equivalent amounts of cash received by the company for the sale of its primary product or service. For oil and gas producing companies, revenue generally refers to the price received for the company's share of oil and gas production.

Cost of goods sold refers to the assets given up or the liabilities incurred by the company in acquiring and disposing of the goods and services represented in revenue. These costs are generally referred to as *product costs* because they define the cost of the product.

Other operating costs refer to those costs incurred in the normal operations and are sometimes lumped with general selling and administrative costs. Included in these costs for oil and gas producing companies are severance taxes and the crude oil windfall profit tax.

Most corporations have substantial current and noncurrent debt. Thus, their *interest expense* is usually quite material to the income statement. For that reason interest expense is usually shown as a separate item.

The *provision for income taxes* is not the same as the income taxes actually paid for the year. A discussion of this is found in

Chapter Nine. For purposes of this chapter, let's simply say that the provision for income taxes is equal to the federal, state and local taxes imposed on the income of the company.

The bottom line, identified as *net income*, is the amount usually referred to as earnings, results of operations or net income. It is the amount usually used in computing the quarterly earnings which are reported to the press and also the amount used to compute earnings per share and the price earnings ratio as reported daily by the *Wall Street Journal.*

Most income statements contain a great deal more information than this basic format. The interested reader is referred to any good intermediate accounting textbook for a thorough discussion of the additional aspects of any of the statements described here.

BASIC INCOME STATEMENT FORMAT

Revenue		XX
Cost and Expenses:		
Cost of Goods Sold	XX	
Other Operations Costs	XX	
General Selling & Administration	XX	
Interest	XX	XX
Net Income Before Taxes		XX
Provision for Income Taxes		XX
Net Income		XX

Figure 1—2

BALANCE SHEET

Refer to Figure 1-3 for the basic format of the corporate balance sheet. The term balance sheet derives from the fact that the total inputs to the company, that is, the contributions made by the creditors and stockholders, should equal the total uses of those inputs, or total assets. To express this another way, *assets* must equal the *liabilities* and *stockholder's equity.*

BASIC BALANCE SHEET FORMAT

Assets

Current Assets:		
Cash	XX	
Accounts Receivable	XX	
Inventories	XX	
Other Current Assets	XX	XX
Non-Current Assets:		
Plant, Property, and Equipment	XX	
Less Accumulated Amortization	-(XX)	
Net Plant, Property, and Equipment	XX	
Other Non-Current Assets	XX	XX
Total Assets		XX

Liabilities and Stockholder's Equities

Current Liabilities:		
Accounts Payable	XX	
Taxes Payable	XX	
Other Current Liabilities	XX	XX
Non-Current Liabilities:		
Bonds and Notes Payable	XX	
Other Non-Current Liabilities	XX	
Deferred Income Taxes	XX	XX
Stockholder's Equity:		
Contributed Capital	XX	
Retained Earnings	XX	XX
Total Liabilities and Stockholders' Equity		XX

Figure 1—3

Note that the assets are divided into two general categories: *current* and *non-current*. Current assets are those assets which are operating cash or will be converted into operating cash within one year or one operating cycle, whichever is longer. These assets are expected to be consumed in normal operations. Inventories will include the costs assigned to any product which has been produced but not yet sold. In the remainder of the Handbook, we will discuss at length the methods by which we determine this product cost.

Non-current assets are those which generally have usefulness or future benefit to the company for more than one accounting period, or for more than one year. The single most important category for oil and gas producing companies is *Plant, Property and Equipment*. There are three general sub-categories:

1. The acquisition cost of *unproved properties*,
2. The acquisition cost of *proved or producing properties*, and
3. The cost of *wells and related equipment and facilities*.

These items are all shown at their gross original cost, even though they may have been consumed through the process of production. The item identified as accumulated amortization represents the amount of reduction of these assets because of production or impairment.

Any non-current asset which has a decline in usefulness to the firm must be subject to some sort of accounting reduction. There are three general types of this reduction:

1. *Depreciation*, which is generally held to be the reduction in usefulness of a tangible fixed asset, such as building or machinery,
2. *Depletion*, which is generally held to be the accounting reduction of a wasting asset such a mineral interest or timber, and
3. *Amortization*, which is generally held to be the accounting reduction associated with an intangible asset, such as a patent, a copyright, or certain leasehold rights.

Since land usually does not decrease in value so far as surface rights are concerned, it generally is not subject to any of these three reductions.

The producing properties of oil and gas companies usually contain all three elements; that is, there are assets which are fixed, there are mineral rights, and there are certain intangibles. Historically, oil and gas producing companies have used the phrase *DD and A*, which refers to depreciation, depletion, and amortization. In referring to the reduction of this orginal cost, the official pronoucements tend to use the single term *amortization* to refer to all three categories. For that reason, amortization will be used throughout this Handbook. However, the reader should be alert to the fact that the term DD and A is widely used in the industry. Most companies will use that expression for their internal financial statements, and then will convert to the single term amortization for external statements.

LIABILITIES

The only liabilities that need specific comment here are those which deal with taxes. *Taxes payable* does not refer to the total tax for the period, but refers instead to taxes which are due but not yet paid and relate to operations that have been completed. For example, a firm may be able to wait 30 days after the end of a period before paying certain payroll or local taxes. This amount due and owing would not be paid until near the last possible date because of the time value of money. That must be shown as an obligation that currently exists. Severance taxes which are assesed against the operations also are included in this amount. Some of these severance taxes may be the expense of the royalty owner or other interests in the property, but were withheld and will be paid by the reporting company. So the taxes payable merely means the company owes these taxes which may or may not be an expense of the company or of the current period.

Deferred income taxes are created because the firm must report on the income statement a provision for income taxes which is usually different from the actual taxes paid. This topic is discussed more thoroughly in Chapter Nine.

STOCKHOLDERS' EQUITY

Contributed capital refers to those funds invested in the corporation by the original stockholders. When a company seeks to raise capital by issuing additional capital stock, the net proceeds from that sale will be placed in this category. Subsequent sales of that stock from one stockholder to another have no bearing on the balance sheet.

Retained earnings does not represent a pool of cash held in the company that is available for stockholders. Rather, it represents the computed net income (which may or may not be in the form of cash, and usually is not), and has not been paid to the stockholders in the form of dividends. The retained earnings for oil and gas producing companies are usually invested in additional acquisition and exploration of properties or development of existing properties. Retained earnings merely represent a claim in addition to the stockholders' contributed capital that the stockholders have on the various assets of the company.

OTHER FINANCIAL STATEMENT INFORMATION

A quick review of the financial statements in Appendix B will clearly demonstrate that a great deal more information is included than the income statement and balance sheet. In Chapter Ten, this additional information is discussed in some detail with specific reference to oil and gas producing companies.

GENERALLY ACCEPTED ACCOUNTING PRINCIPLES

Previous reference was made to generally accepted accounting principles. There are seven such principles which need to be discussed in more detail here, but first a word of caution. Contemporary financial accounting principles have been developed over several decades. There have been hundreds of official pronouncements by the various bodies and countless unofficial documents which define generally accepted accounting principles. The former Accounting Principles Board referred to generally accepted accounting principles as those accounting principles which are generally accepted.[2] This may at first appear to be circular reasoning, but what the Board was saying is that there are a large number of practices which are followed by most

[2]"*Generally accepted accounting principles* therefore is a technical term in financial accounting. Generally accepted accounting principles encompass the conventions, rules, and procedures necessary to define accepted accounting practice at a particular time. The standard of generally accepted accounting principles includes not only broad guidelines of general application, but also detailed practices and procedures." Source: Accounting Principles Board Statement No. 4, paragraph 138.

companies but are not dealt with by specific pronouncements. Since these are generally practiced, they should be followed by companies that claim to follow the generally accepted accounting practices of their industry. This section will discuss briefly the *historical cost concept, comparability* and *consistency, the lower of the cost or market concept, amortization, materiality, capitalization*, and certain terminology concepts.

HISTORICAL COST

One of the most frequently held misconceptions about financial statements is that the dollar amounts on the balance sheet represent the value of the assets; that is, if a company shows plant, property, and equipment of $50,000,000 on the balance sheet, that company has plant property worth $50,000,000. Nothing could be further from the truth. One of the most pervasive of all accounting principles is that amounts are put into the accounts at their historical cost, or the amount that the company gave up in order to acquire those assets. This may or may not be close to today's economic reality. Consider, for example, the company that builds an office building in a metropolitan area, and then shows that amount on the balance sheet at the cost given up for the building. Ten years later, because real estate values have continued to climb, the actual value of the building has substantially increased. However, for accounting purposes, the company has depreciated the building so that its carrying value on the balance sheet is actually lower than its original cost. In this case we have costs which are substantially lower than the value.

Consider the unique case of oil and gas reserves. The balance sheet will disclose the cost of finding and developing the reserves. This amount will clearly be substantially different, usually less, than the actual economic value of those reserves. For this reason we must be careful not to interpret any account balance as being equal to the value of any item.

The S.E.C. has attempted to overcome this problem in the case of oil and gas reserves by requiring companies to disclose information based, not on historical cost of the property, but rather on a measure of the future cash expected to be received. This controversial accounting technique is called *Reserved Recognition Accounting* (RRA), and while it will not be used in the primary financial statements, it is currently required as supplementary disclosure by oil and gas producing companies who report to the S.E.C. A thorough discussion of RRA is found in Chapter Ten.

COMPARABILITY AND CONSISTENCY

Another of the most basic and pervasive generally accepted accounting principles is that of comparability and consistency. If an investor wishes to invest in the stocks or bonds of an oil and gas producing company, one of his best sources of information is the financial statements of that company. Certainly the investor would like to be able to review the statements of several companies in order to make an intelligent decision. The comparison of those statements can be meaningful only if all statements were prepared using the same set of rules. Consider how impossible it would be to compare statements of a company prepared on the basis of historical cost to a company's statements prepared on the basis of Reserve Recognition Accounting. For that reason, companies must carefully disclose any deviations they have from generally accepted accounting principles and when various options are available, such as valuing inventory under several different methods, the companies must disclose which option they have selected.

It is also assumed that the company will be consistent in an application of alternative solutions. For example, if a company values its inventory on the last-in-first-out basis in 1980, it is also assumed (unless there is a notice to the contrary) that the 1981 inventory is valued on the last-in-first-out basis, rather than the first-in-first-out basis.

A most perplexing problem exists in the oil and gas industry in that two widely different methods are used throughout the industry for valuing the cost of acquiring, exploring for, and developing oil and gas reserves. One method, refered to as *successful efforts*, and the other method, referred to as *full cost*, are used by almost equal numbers of oil and gas producing companies. A thorough discussion of the difference between these methods is found in Chapter Four.

LOWER OF COST OR MARKET

Financial statements tend to be *conservative*; that is, if a company has two options under which to report an asset it should usually choose the option which tends to report the lower amount for the asset. If a company has two options under which to report an expense, it should generally report the higher value for that expense. With this concept in mind, the profession has developed a technique for valuing assets which is referred to as the lower of cost or market. Simply stated, the company must compare the historical cost of the assets, minus any accumulated amortization, to the net realizable value for that asset. If the realizable value is less than the historical cost, the value of the asset is written down to market value. If the market value is greater than the historical cost, the asset continues to be valued at its original cost, less accumulated amortization, if any.

Consider the case of a company which has only one oil and gas lease. If that lease is shown to have proved reserves, the probability is that its net realizable value or its market value is much greater than the total amount paid to acquire, explore and develop the lease. Therefore, that lease would be carried at its historical cost. On the other hand, if that lease is shown to have no reserves, then its market value is probably much less than the amounts spent on the property; therefore its cost would be written down to its market value. If the lease is impaired,[3] there is a strong possibility that the market value is zero. In that case, the lease would be written down to a zero carrying value and none of the historical cost would be carried forward on the balance sheet. All the costs would be transferred to an expense on the income statement.

AMORTIZATION

We have previously discussed the terminology for amortization. Let's now briefly look at the theory behind it. An asset which has a long but not indefinite life, will benefit the firm over several accounting periods. The easiest example is to consider a factory building. Perhaps the building has a life expectancy of 40 years. The original cost of that building must be charged against revenue in order to determine net income. If we were to produce only one income statement over the 40-year period, we would simply charge all of the cost of that building against one expense entry. However, we normally prepare financial statements at least annually, and usually quarterly, so that it is necessary to take a portion of the cost of that building and charge it against the revenue

[3]See Chapter Five for a discussion of impairment.

for that shorter period. We have several methods available to determine what should be charged each accounting period. For an asset such as a building we may use a *straight line method*; that is, each year we charge one-fortieth of the cost of the building against revenue. Or we may use one of the *accelerated depreciation* methods, whereby larger amounts are deducted in earlier years of the life of the asset and smaller amounts in later years. Under any method, the total amount charged over the life of the asset should be the same.

For oil and gas producing properties, however, we use a different method. This is referred to as the *units of production* method. In this case we will determine the total cost of the revenues and the total reserves available for production and calculate a rate per barrel. This rate will then be applied to the actual production that we have for the period. In this way each barrel will theoretically bear the same cost. There are many adjustments and complicating factors in computing amortization, and they are dealt with in detail in Chapter Eight.

CAPITALIZATION

If an expenditure is *capitalized*, it will have benefit to the firm for more than the current accounting period. Its cost is recorded as an asset and is written off to expense as its value to the firm decreases.

On the other hand, expenditures that do not have future benefit to the firm, or are not material, are *expensed* as incurred and never appear on the balance sheet as an asset.

MATERIALITY

If an amount is not large enough to affect the financial statements, it is said to be not material. Items which are not material need not be accounted for under generally accepted accounting principles.

The problem, of course, is to define the level of materiality. That level is generally left to the decision of the preparer and his auditor. A rule of thumb is that, if a decision by the statement user could be affected by the disclosure of an item, that item is material.

TERMINOLOGY

One of the most confusing terms used in oil and gas accounting is the term *reserve*. Reserve from an engineering standpoint refers to the product of crude oil or natural gas underground. The term reserve is also used in accounting as a *contra account* or a reduction of another account. For example, accumulated amortization, which is subtracted from plant, property, and equipment, is sometimes referred to as Reserve for DD&A. This means that it is an amount that is subtracted from the primary account. The former Committee on Accounting Practices recognized the complications with the term reserve and limited the use of the word reserve in financial statements to appropriations of retained earnings.[4] That is, the term reserve could not be used to refer to accumulated amortization or other similar accounts. This was done because many readers may have assumed that the reserves referred to a pool of cash or, in the case of oil and gas companies, to product reserves in the ground.

Most companies continue to use the expression Reserve for DD&A in their internal accounting statements. Some companies also use it in their external financial statements, although technically it is improper to do so. The reader of any financial statement should be alert to this confusing use of the term reserve.

[4]"The generally accepted meaning of the term reserve corresponds fairly closely to the accounting usage which indicates an amount of unidentified or unsegregated assets held or retained for a specific purpose. This is the use to which the committee feels it should be restricted." Source: Accounting Terminology Bulletin No. 1, paragraph 67.

Chapter 2

Unique Characteristics of Oil and Gas Reserves and Production

This chapter will introduce you to the oil and gas industry, the unique characteristics of oil and gas reserves in the ground, and the production of those reserves. An understanding of these characteristics and the terms associated with them is necessary in order to fully appreciate the accounting complexities we will deal with in later chapters.

GENERAL INDUSTRY CHARACTERISTICS

The oil and gas industry is the most dominant industry in the United States' economy. (Refer to Table 2-1.) It is clear that 13 of the 20 largest corporations in the United States are oil and gas producing companies, and have a combined asset base of $263 billion dollars and sales of $421 billion dollars. When we add to these top corporations the 1,100 or so other oil and gas producing companies that publish financial statements and the thousands of smaller independent, closely held oil and gas producing companies, we realize that the oil and gas industry dominates our economy above any other industry.

Many people frequently think of the automotive industry as the most dominant in our economy. However, notice that only two of the top 20 corporations are automotive companies and there are only two other significant automotive companies in the United States.

It is clear that any significant change in the oil and gas industry will have a profound effect on our economy. By the same token, any significant change in the accounting requirements for oil and gas producing companies will affect the accounting profession and the financial community.

The second most significant characteristic of the oil and gas industry is the large risk taken by oil and gas producers. A single exploratory well drilled in a hostile environment can cost many million dollars. There is a high probability that these exploratory

THE 20 LARGEST US INDUSTRIAL CORPORATIONS

	SALES		ASSETS		NET INCOME	
	($MM)	Rank	($MM)	Rank	($MM)	Rank
*Exxon	103,143	1	56,577	1	5,650	1
*Mobil	59,510	2	32,705	3	3,272	3
General Motors	57,729	3	34,581	2	(763)	490
*Texaco	51,196	4	26,430	5	2,643	4
*Standard Oil (Calif)	40,479	5	22,162	7	2,401	5
Ford Motor	37,086	6	24,348	6	(1,543)	491
*Gulf Oil	26,483	7	18,638	9	1,407	11
IBM	26,213	8	26,703	4	3,562	2
*Standard Oil (Ind.)	26,133	9	20,167	8	1,915	6
General Electric	24,595	10	18,511	10	1,514	10
*Atlantic Richfield	23,744	11	16,605	12	1,651	8
*Shell Oil	19,830	12	17,615	11	1,542	9
IT&T	18,530	13	15,417	13	894	15
*Conoco	18,325	14	11,036	18	1,026	14
du Pont	13,652	15	9,560	22	716	21
*Phillips Petroleum	13,377	16	9,844	20	1,070	13
*Tenneco	13,226	17	13,853	14	726	19
*Sun	12,945	18	10,955	19	723	20
U.S. Steel	12,492	19	11,748	16	505	33
*Occidental Petroleum	12,476	20	6,630	33	711	22

**Indicates company whose dominant line of business is oil and gas production.*
Source: Fortune, May 4, 1981.

Table 2—1

wells will be dry holes. Hundreds of millions of dollars can be spent exploring a region, and that region may ultimately prove to have no producable reserves.

Even on a smaller scale, an independent oil and gas producing company will invest a very large percent of its discretionary capital in drilling each exploratory well. The company that suffers a successive long string of dry holes and impairment of unproved properties will have a very difficult time staying in the capital market as a going concern. Offsetting this risk is the potential for a very high return. As the current price of crude oil is over $30 dollars per barrel, the potential financial return for large discoveries of proved reserves is substantial.

The third major industry characteristic is that of technical complexity. There is perhaps no other industry in the United States that deals with so many highly complex engineering problems just to produce its product. Not only must oil and gas producing companies face substantial technical problems in discovering the product, they must also overcome substantial technical problems in producing and selling the product on a competitive basis.

The final general industry characteristic is that of the level of regulation for the industry. Oil and gas producing companies are regulated at the local, state, and federal levels. It is not an understatement to note that oil and gas producing companies are among the most regulated in the United States today. Companies are regulated by where they may drill, when they may drill, how deep they may drill, how fast they may produce the product, to whom they may sell the product, what they may charge for the product, and how they can ship the product. In short, there is no phase of the industry that is not subject to substantial regulation. We have, in recent months, undergone a deregulation of crude oil prices. However it should be noted that most natural gas is still price regulated, and that the production, transportation, and refining of crude oil is still heavily regulated.

In summary, the oil and gas industry dominates our economy but does so in an environment that is highly complex and requires a great deal of risk on the part of the companies. This is further accomplished within a regulatory environment that is more severe than in most other industries.

FORMATION OF OIL AND GAS RESERVES

Oil and gas reserves began as sea life more than a hundred million years ago. These sea plants and animals died, fell to the ocean floor and were covered by successive layers of sediment. Over millions of years various types of layers of rock were created in the bottom of the ocean. Inside this rock, through a process of heat, pressure, bacteria and enzymes, the former sea plants and animals were converted into the hydrocarbons, crude oil and natural gas.

Perhaps the most common misconception about oil and gas reserves is that they exist under the ground in large caverns or pools or rivers. Nothing could be further from the truth. In fact, crude oil and natural gas are found inside rock, primarily *sand-*

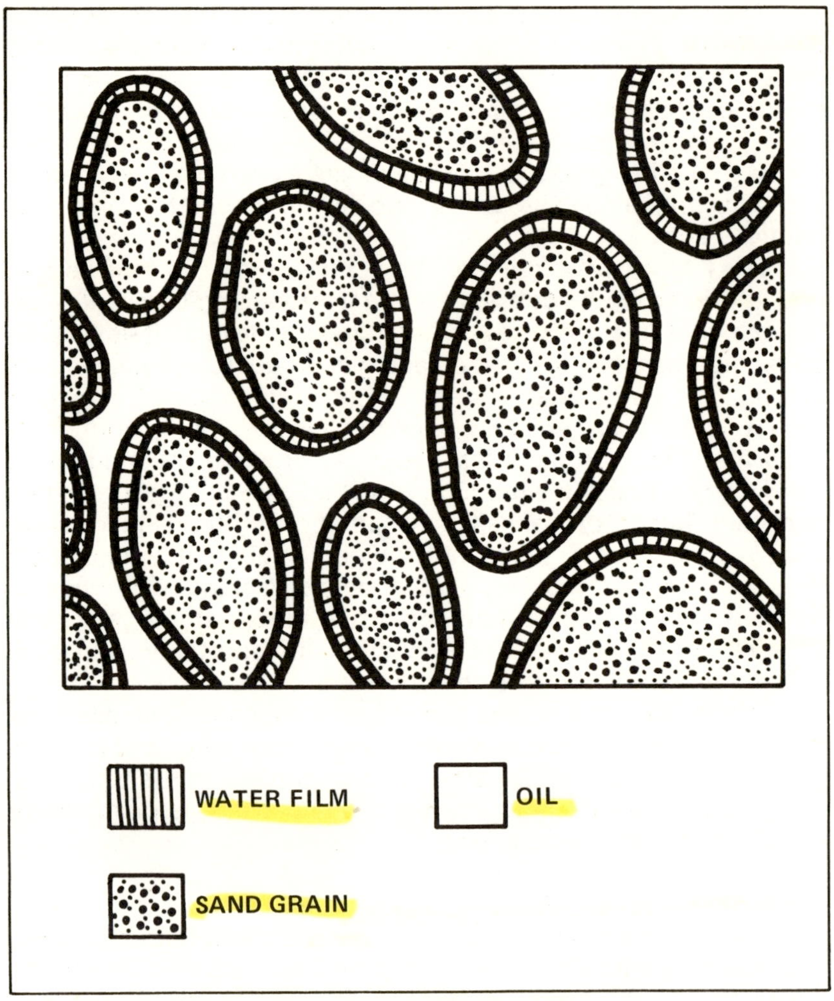

Figure 2–2 Sketch of sandstone showing pores and water film around grains.

stone and *limestone*. As illustrated in Figure 2-2 the liquid crude oil and the gaseous natural gas are found in between the grains of sand, or the pores, of the sandstone. The percent of the total volume of this rock which is represented by empty space between the pores is refered to as *porosity*. It is not uncommon for a sandstone or a limestone to have as little as 10 or 15 percent porosity.

That is, for every one hundred cubic feet of rock there will be 10 or 15 cubic feet of space between microscopic grains. It is in this porosity that we find the crude oil and natural gas.

The product generally occurs underground anywhere from a few thousand feet to as deep as 20 thousand feet or more. At these depths, there exists a high level of pressure. This pressure forces the product upward through the rock and the product would continue its journey if it were not stopped by layers of other rock, for example, *shale*, which is much harder than the sandstone and limestone. That is, shale has a very low *permeability* and the product cannot pass through it. (See Figure 2-3.) It is in natural forma-

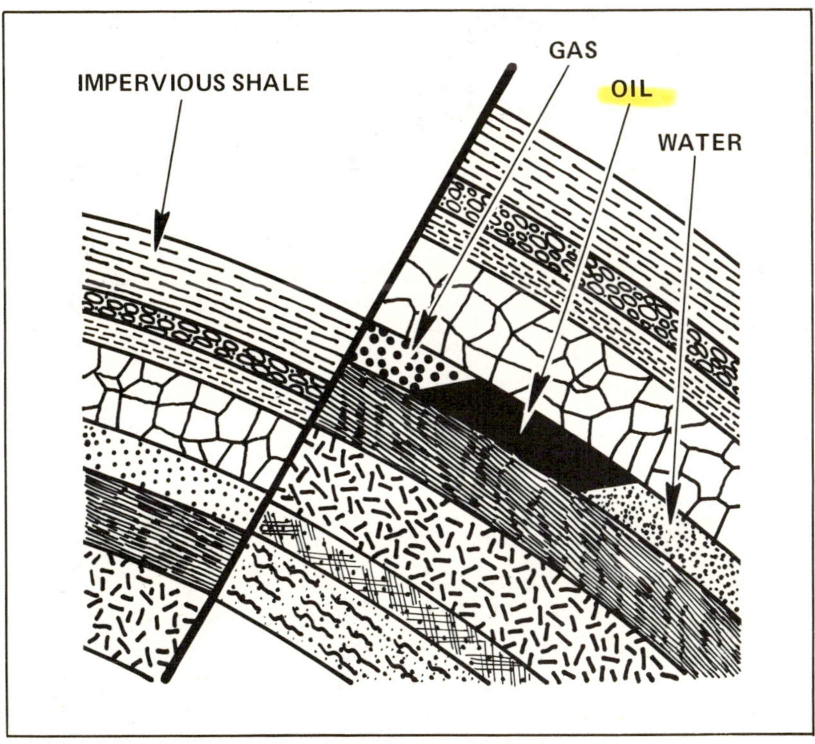

Figure 2—3 Gas and oil are trapped in a fault trap—a reservoir resulting from normal faulting or offsetting of strata. The block on the right has moved up from the block on the left, moving impervious shale opposite the hydrocarbon-bearing formation.

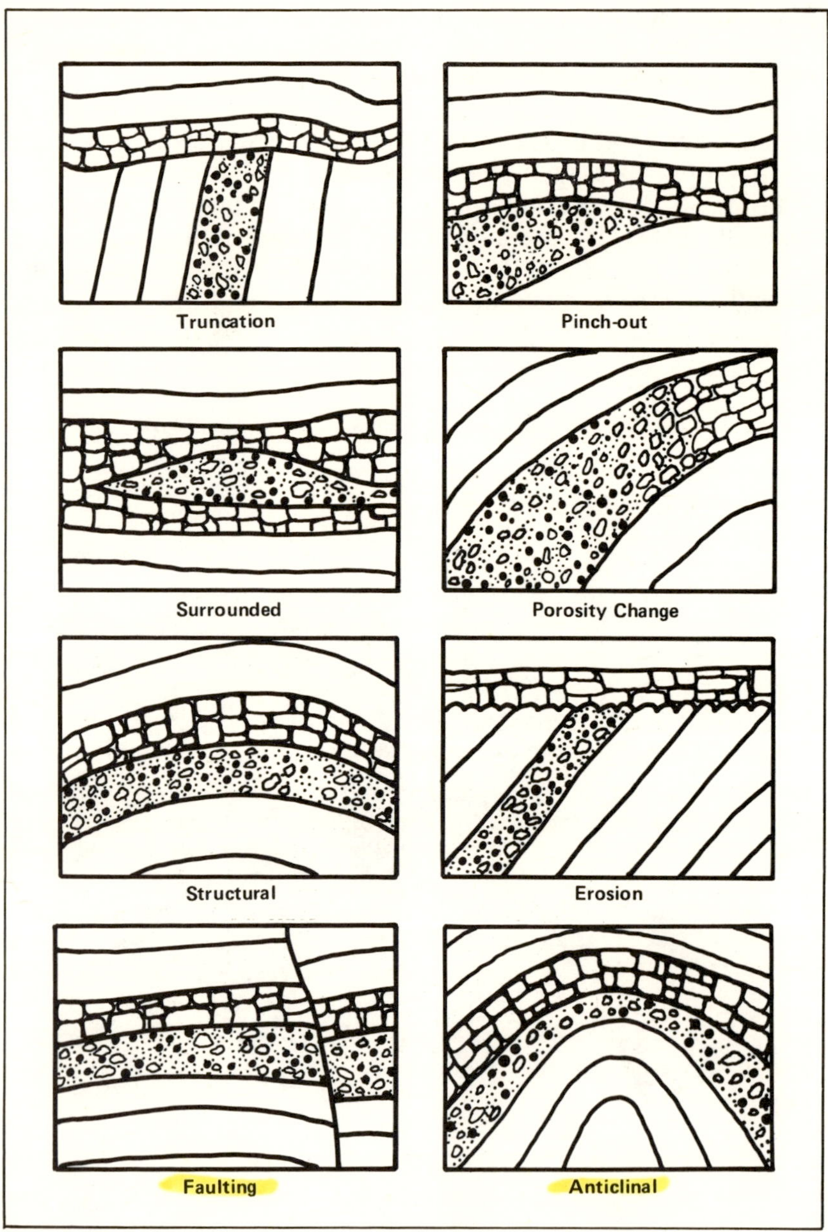
Truncation
Pinch-out
Surrounded
Porosity Change
Structural
Erosion
Faulting
Anticlinal

Figure 2–4

tions then where sandstone and limestone are trapped by shale or other impervious rock that we find crude oil and natural gas. Figure 2-4 illustrates some of these typical formations.

DISCOVERY OF RESERVES

In order to find those formations where oil and gas reserves are likely, exploratory *seismic studies* are conducted, as illustrated in Figure 2-5. In the seismic studies, an explosion or a dropped weight sends a shock wave down through the various layers of rock. The waves bounce off the rock back up to the surface where they are recorded. Careful analysis of this data produces a *contour map*, as illustrated in Figure 2-6.

The seismic studies and contour maps do not discover oil and gas reserves. They only reveal formations, such as anticlines or faults, that may contain oil and gas reserves. The only certain way to determine whether reserves exist is to drill into the formation. The seismic studies and the contour maps do indicate the favorable locations where drilling should occur.

Unfortunately, most exploratory wells are not successful. Over the last 20 years, on a national average, only one out of every nine exploratory wells has found new reserves. Some enterprises have been more successful and some have been less, but it is clear that most drilling efforts of a exploratory nature will not be successful. Add to this the high cost of drilling the well, (the current average well cost in the United States today is somewhat above $2 million per exploratory well) and we can see the inordinate risk taken by oil and gas producing companies.

DEFINITION OF PROVED RESERVES

We shall see later that whether or not a well has discovered *proved reserves* is not only a critical economic event for the company, but is very important in many accounting matters. For accounting purposes we use the following definition of proved reserves:

> *Proved oil and gas reserves.* Proved oil and gas reserves are the estimated quantities of crude oil and natural gas, and natural gas liquids which geological and engineering data demonstrate with reasonable certainty to be recoverable in future years from known reservoirs under existing economic

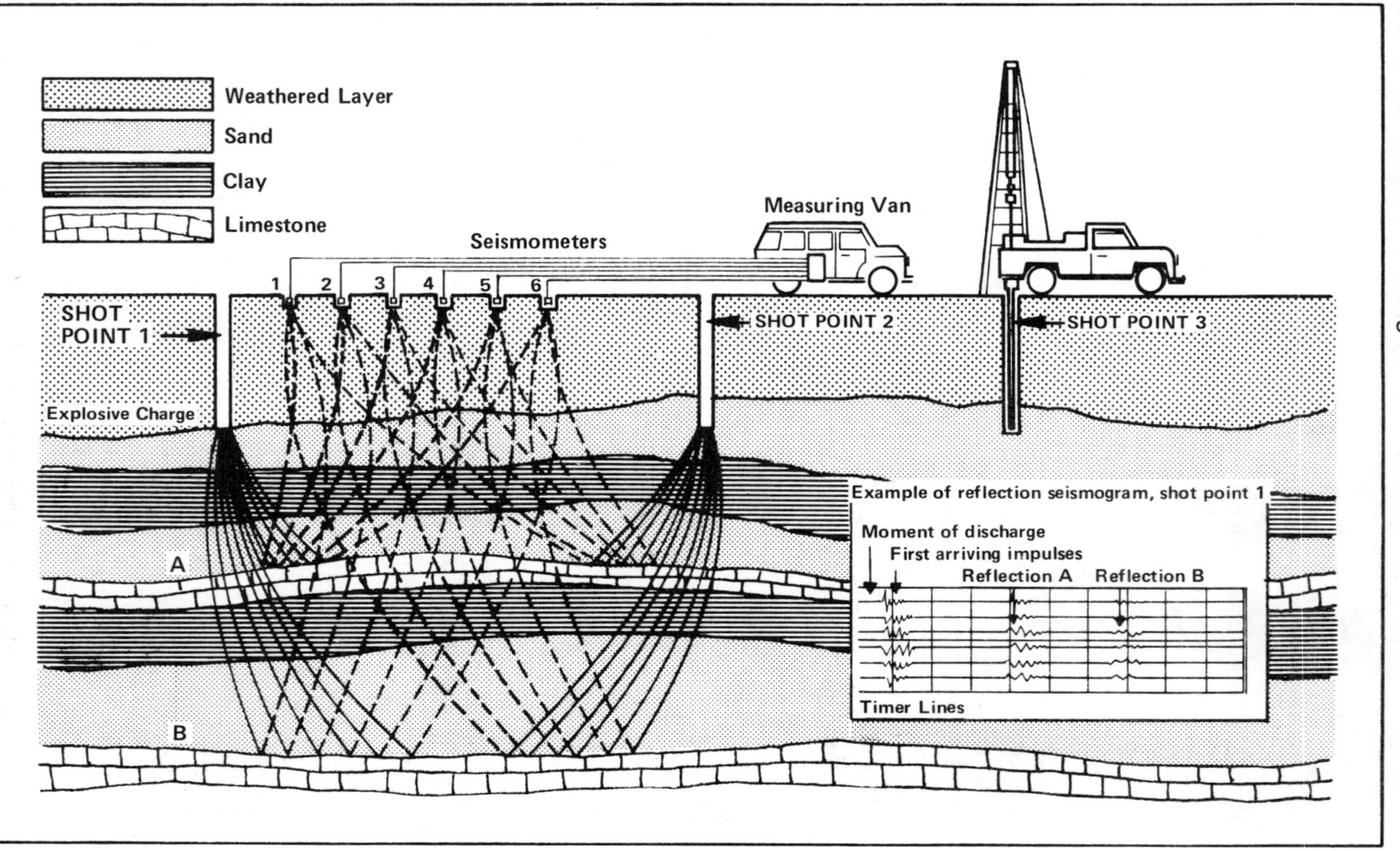

Weathered Layer
Sand
Clay
Limestone
Measuring Van
Seismometers
1
2
3
4
5
6
SHOT POINT 1
SHOT POINT 2
SHOT POINT 3
Explosive Charge
A
B
Example of reflection seismogram, shot point 1
Moment of discharge
First arriving impulses
Reflection A
Reflection B
Timer Lines

Figure 2—5

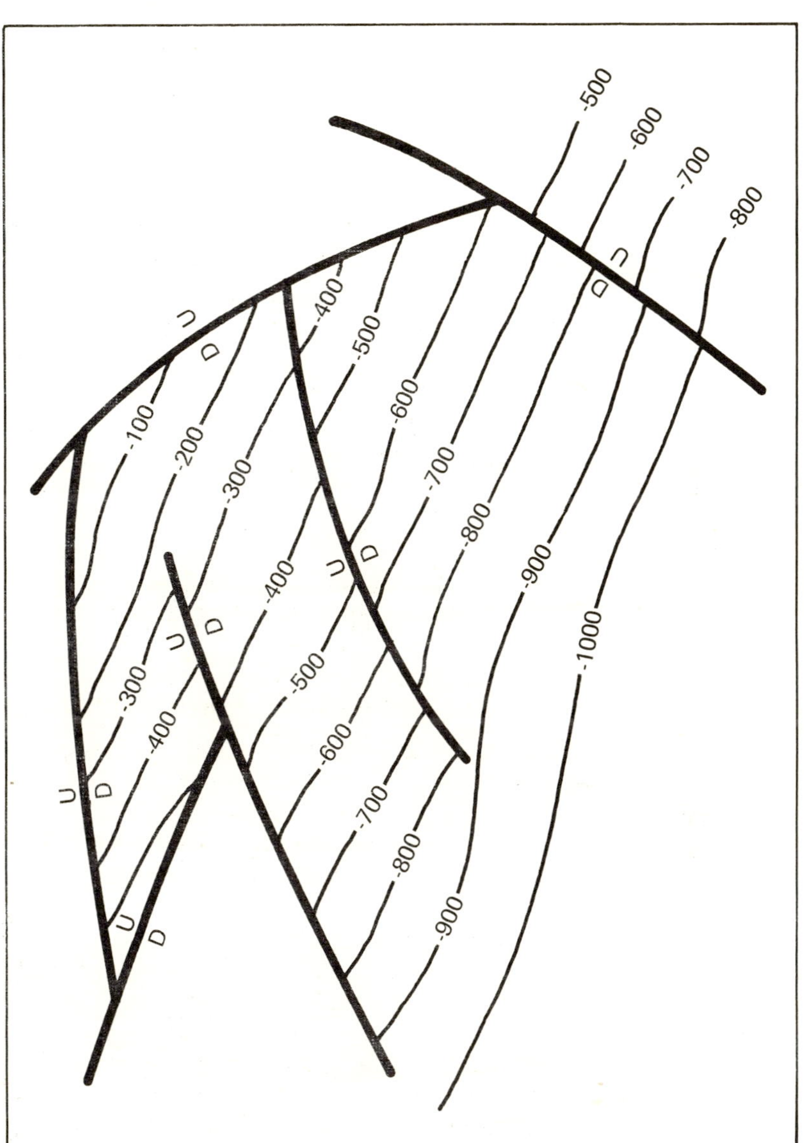

Figure 2–6 Normal fault contour map.

and operating conditions, i.e., prices and costs as of the date the estimate is made. Prices include consideration of changes in existing prices provided only by contractual arrangements, but not on escalations based upon future conditions.

1. Reservoirs are considered proved if economic producibility is supported by either actual production or conclusive formation test. The area of a reservoir considered proved includes (a) that portion delineated by drilling and defined by gas-oil and/or oil-water contacts, if any, and (b) the immediately adjoining portions not yet drilled, but which can be reasonably judged as economically productive on the basis of available geological and engineering data. In the absence of information on fluid contacts, the lowest known structural occurrence of hydrocarbons controls the lower proved limit of the reservoir.

2. Reserves which can be produced economically through application of improved recovery techniques (such as fluid injection) are included in the "proved" classification when successful testing by a pilot project, or the operation of an installed program in the reservoir, provides support for the engineering analysis on which the project or program was based.

3. Estimates of proved reserves do not include the following: (a) oil that may become available from known reservoirs but is classified separately as "indicated additional reserves"; (b) crude oil, natural gas, and natural gas liquids, the recovery of which is subject to reasonable doubt because of uncertainty as to geology, reservoir characteristics, or economic factors; (c) crude oil, natural gas, and natural gas liquids, that may occur in undrilled prospects; and (d) crude oil, natural gas, and natural gas liquids, that may be recovered from oil shales, coal, gilsonite and other such sources.

Proved developed oil and gas reserves. Proved developed oil and gas reserves are reserves that can be expected to be recovered through existing wells with existing equipment and operating methods. Additional oil and gas expected to be obtained through the application of fluid injection or other improved recovery techniques for supplementing the natural forces and mechanisms of primary recovery should be included as "proved developed reserves" only after testing by a pilot project or after the operation of an installed program has confirmed through production response that increased recovery will be achieved.

Proved undeveloped reserves. Proved undeveloped oil and gas reserves are reserves that are expected to be recovered from new wells on undrilled acreage, or from existing wells where a relatively major expenditure is required for recompletion. Reserves on undrilled acreage shall be limited to those drilling units offsetting productive units that are reasonably certain of production when drilled. Proved reserves for other undrilled units can be claimed only where it can be demonstrated with certainty that there is continuity of production from the existing productive formation. Under no circumstances should estimates for proved undeveloped reserves be attributable to any acreage for which an application of fluid injection or other improved recovery technique is contemplated, unless such techniques have been proved effective by actual tests in the area and in the same reservoir.[1]

Thus, we can paraphrase that proved reserves are those reserves which are economically feasible to develop and produce in today's economy and with today's technology. Reserves which were not feasible four or five years ago when the price of crude oil was $14 a barrel may be feasible in today's environment of over $30 a barrel of oil.

Consider some other events that could change the amount of proved reserves. Perhaps we have discovered oil which is very thick, and does not easily flow through the rock, but a change in the technology of production now makes these reserves economically feasible. Or consider the effects of the 1980 Crude Oil Windfall Profit Tax. Certainly there are reserves which were

[1]Financial Accounting Standards Board, "Statement of Financial Accounting Standards No. 25," paragraph 34.

not feasible with this tax, but would have been produced had there been no tax.

The point to make here is that the definition of a particular reservoir as proved or unproved is as much determined by economic, political, and technological events, as it is by the actual discovery of hydrocarbons under the ground. We shall address the accounting implications of this definition in later chapters.

TESTING OF A HYDROCARBON DISCOVERY

When an exploratory well shows that crude oil or natural gas exists, the exploring company must follow a series of tests to determine whether or not enough reserves exist to justify completion and development of that well.

First the well is *logged*; that is, a device is run through the bore to determine the electrical resistance and potential of the formation. See, for example, the well log in Figure 2-7. Through the log of the well the number of feet of pay zone in a particular formation can be estimated. This is the number of feet of crude oil or natural gas bearing rock that exists in the formation. By comparing the contour map and the results of the well logging, we can estimate the total volume of the formation that can produce hydrocarbons.

We then take a *core sample*, illustrated in Figure 2-8, and by examining this sample we can estimate both the porosity and the permeability of the formation.

Finally, we allow the well to flow for a brief period of time and measure the pressure as well as the volume of water and other impurities in the product.

By examining all this data we are able to make an educated guess on whether or not reserves are sufficient to justify the additional expenses to be incurred in completing and developing the property. It should be noted here that this evaluation of the well is an inexact art at best. Different but capable engineers given the same data will frequently come up with different recommendations. It is not unusual for a well to be completed and then later to discover that the reserves were not really adequate to justify that expenditure. On the other hand, there are certain formations that have not been completed that really did have adequate reserves.

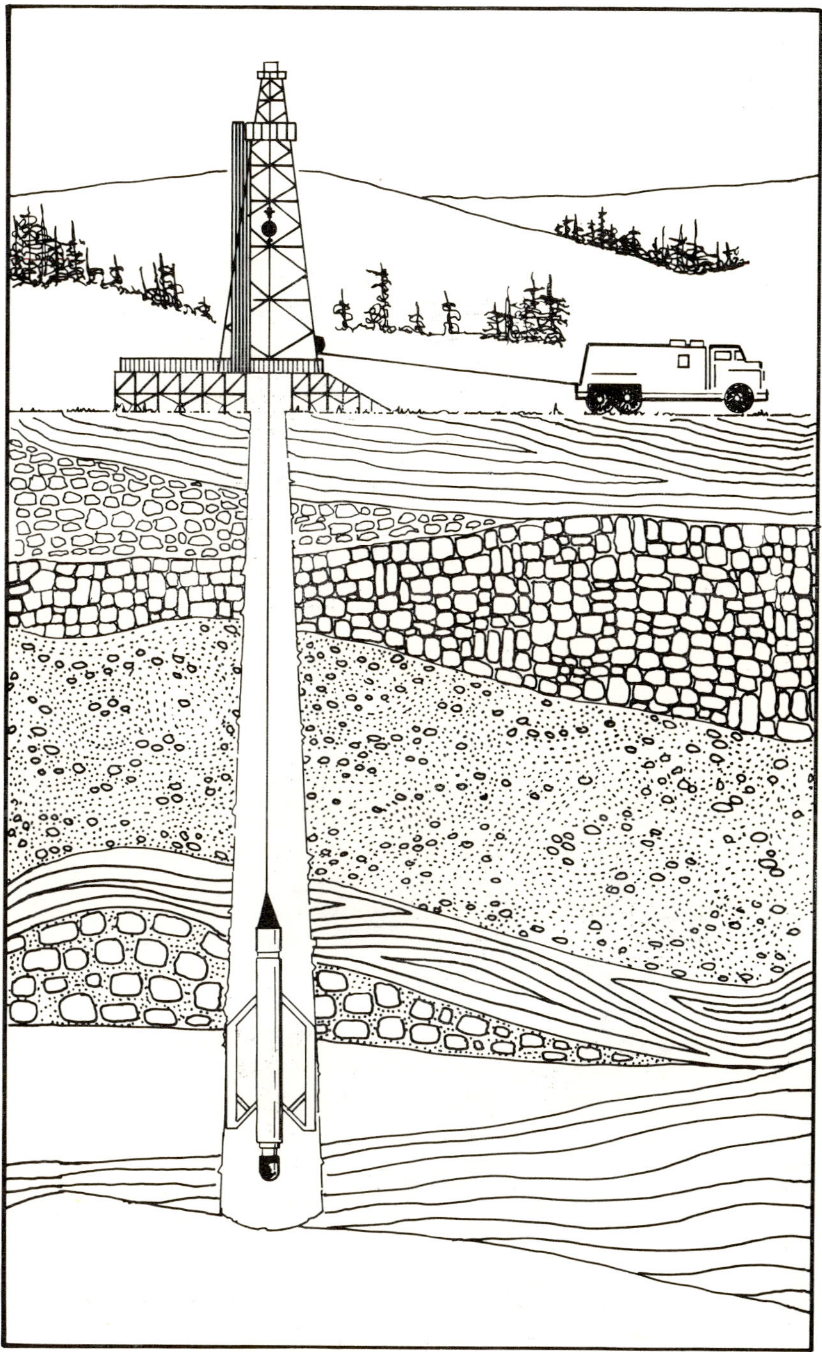

Figure 2—7 A well-logging truck is brought to the well site while the drill string is tripped out.

Figure 2—8 Core Samples

THE DEVELOPMENT PROCESS

The discovery well is completed; the formation is first *cased and cemented*, as illustrated in Figure 2-9. This isolates the formation and contains the pressure within the formation.

This isolated formation is then *perforated* by a special device that electronically fires projectiles through the casing, cement, and into the rock, as illustrated in Figure 2-10. This perforation then allows a sieve-like entry into the well bore. The well is sometimes *fractured*; hydraulic fluid or other materials are forced down into the well bore under very high pressure. This fluid enters the perforation and goes into the rock where it cracks the rock and increases the permeability of the formation.

If there is adequate pressure in the crude oil formation (and there usually is in a new discovery) only a christmas tree, as illustrated in Figure 2-11, is placed on the surface of the well. The natural pressure of the formation will then force the crude oil to the top where it is treated and sold. After a period of time the natural pressure decreases and the well must be produced by means of artificial lift, as illustrated in Figure 2-12. Natural gas cannot be lifted from the ground by means of artificial lift, and must flow under its own pressure. All natural gas wells will have a christmas tree at the surface.

A series of additional or *development wells* are usually drilled to produce all portions of the reservoir.

CRUDE OIL SURFACE TREATMENT

Crude oil exists in the underground rock formation as a combination of products. In addition to the crude oil itself, there are variable quantities of water and natural gas. If a formation is primarily producing oil the natural gas which is produced along with it is referred to as *associated natural gas*, because it is associated with the production of the crude oil.

At the surface these three products must be separated. Some of the natural gas will fall out naturally as the product is brought to the surface and undergoes changes of both temperature and pressure. This gas which comes out freely at the surface is referred to as casinghead gas, because it can be recovered at the top of the surface casing or the casinghead of the well. Other natural gas is dissolved in the crude oil and along with the water must be removed in a device called a *separator*. A common form of separator is a *heater treater*, as illustrated in Figure 2-13. In this

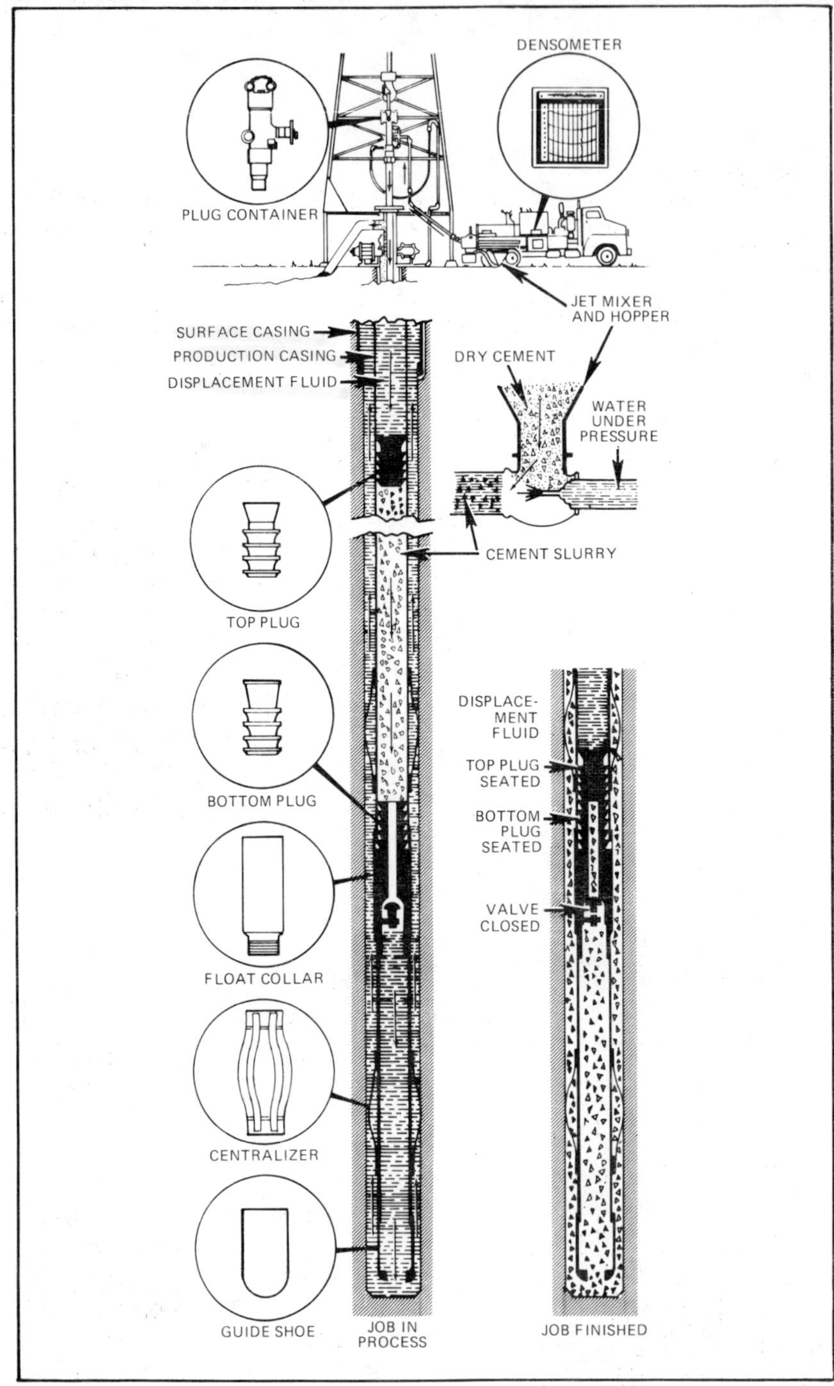

Figure 2–9 Casing and Cementing

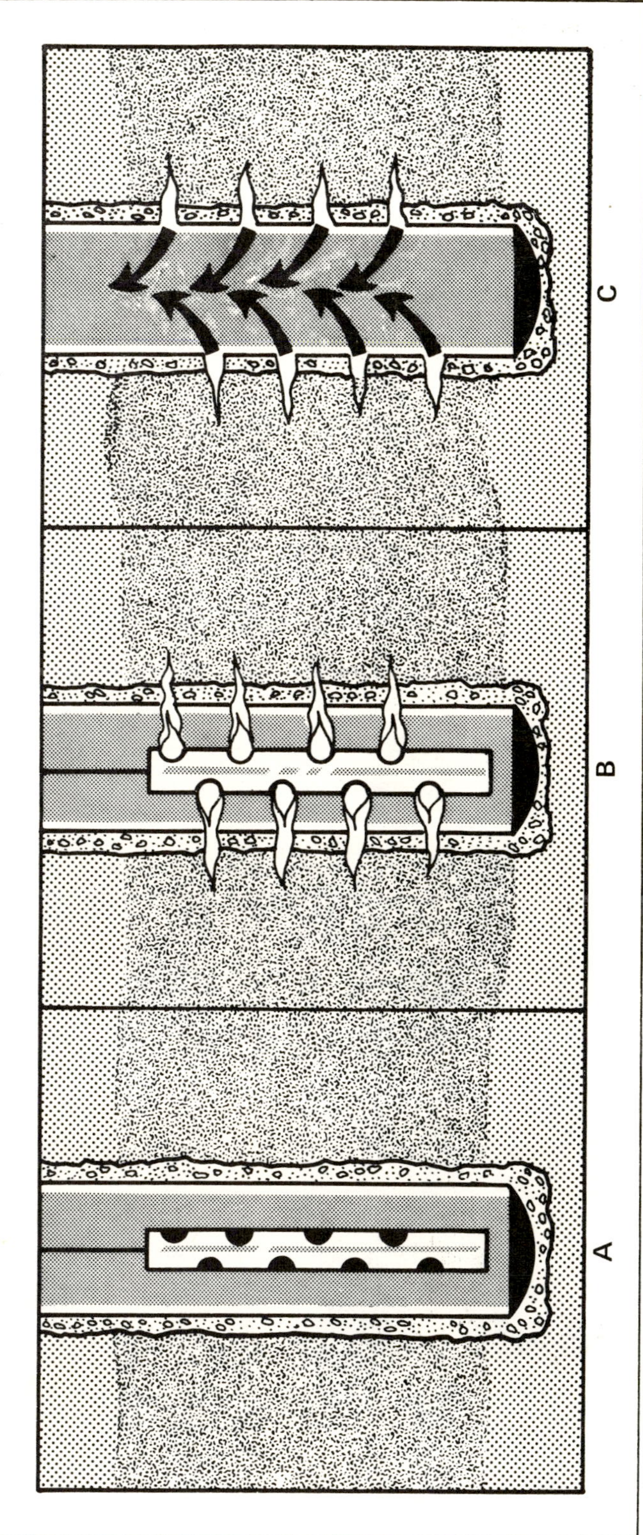

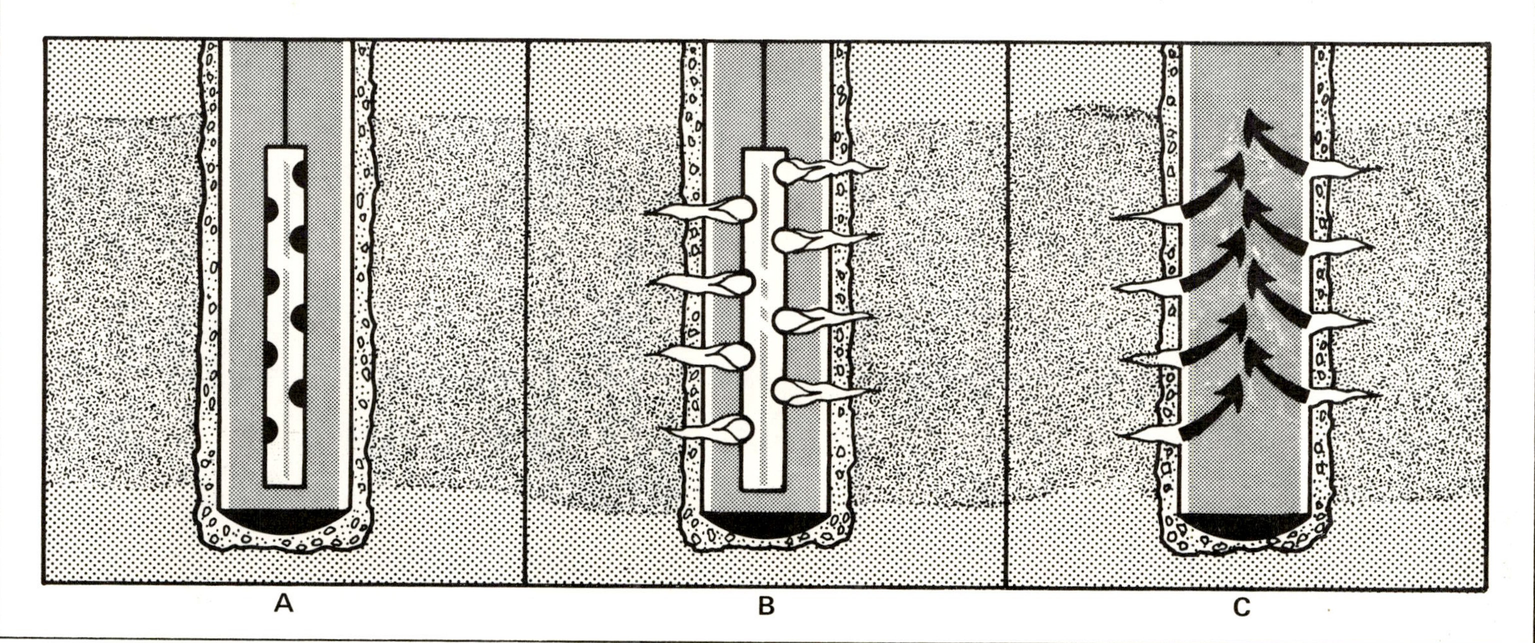

Figure 2–10 The perforating gun (A) is lowered into the wellbore to a depth opposite the producing zone. The shaped charges are fired (B), resulting in several perforations (C) that allow reservoir fluids to flow into the wellbore.

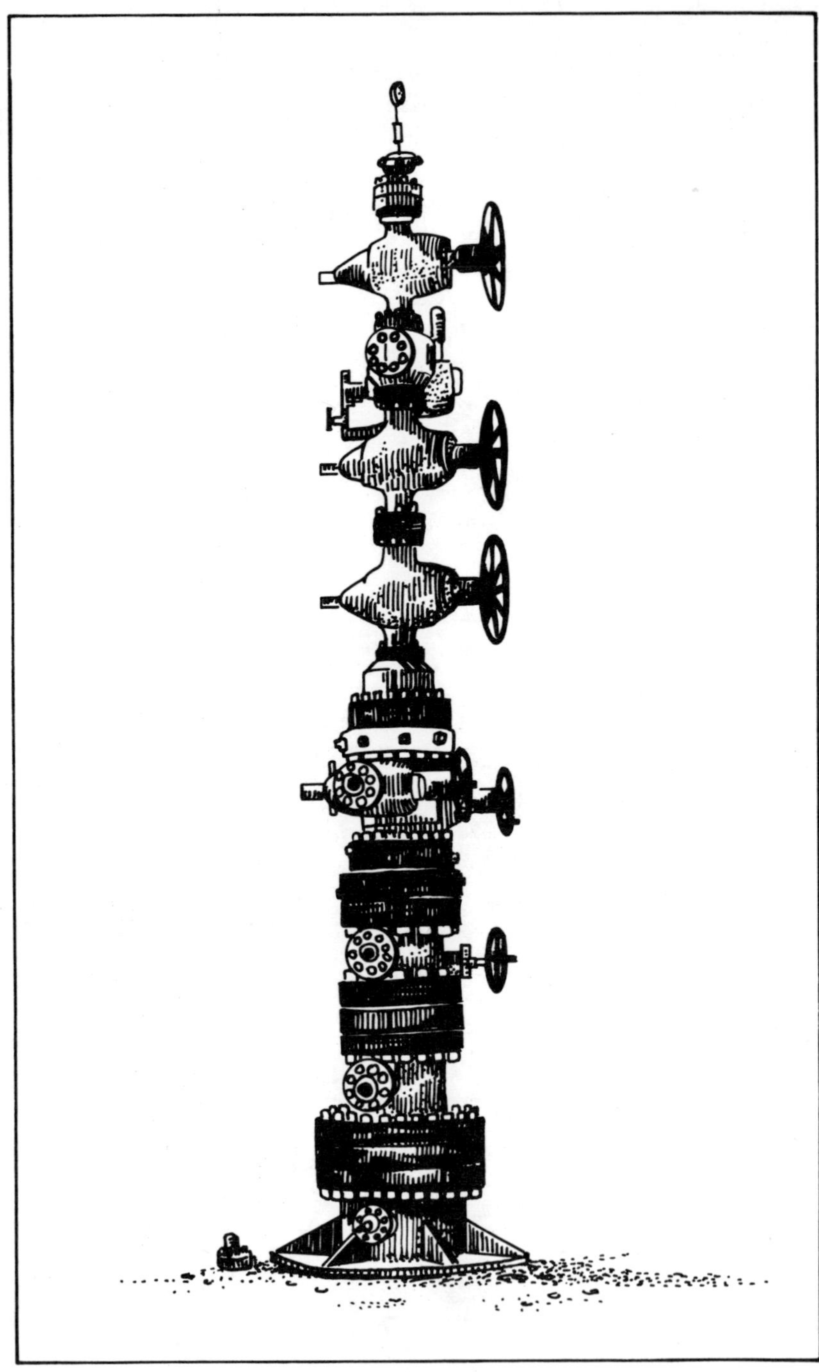

Figure 2–11 Christmas Tree

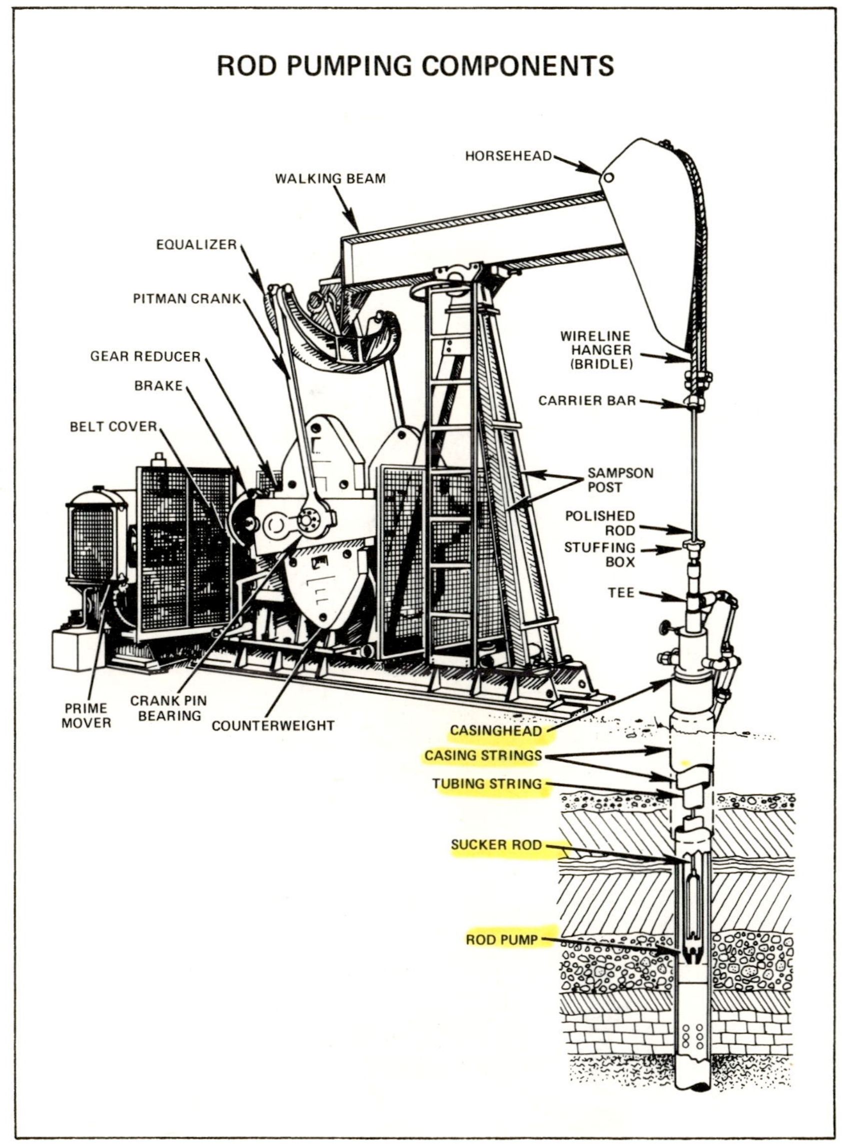
ROD PUMPING COMPONENTS
HORSEHEAD
WALKING BEAM
EQUALIZER
PITMAN CRANK
GEAR REDUCER
BRAKE
BELT COVER
WIRELINE HANGER (BRIDLE)
CARRIER BAR
SAMPSON POST
POLISHED ROD
STUFFING BOX
TEE
PRIME MOVER
CRANK PIN BEARING
COUNTERWEIGHT
CASINGHEAD
CASING STRINGS
TUBING STRING
SUCKER ROD
ROD PUMP

Figure 2–12

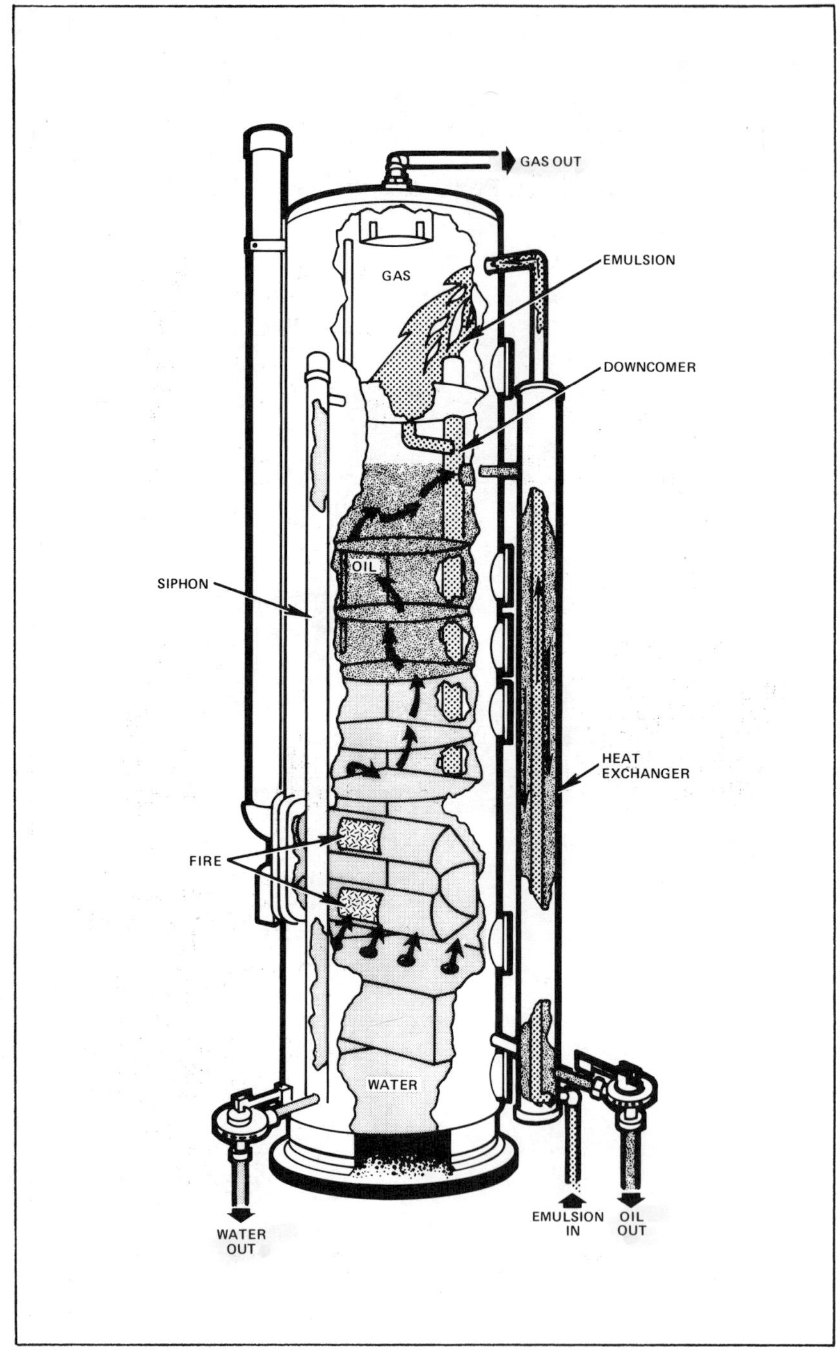

Figure 2–13 Cutaway view of a vertical heater-treater.

device the products are heated and the dissolved natural gas boils off the top where it is recovered. The water, which is heavier, falls out at the bottom and is referred to as *basic sediment and water*, or BS and W. The crude oil in the middle is then placed in a *holding tank* where it may be left for a period of time to allow additional BS and W to settle to the bottom of the tank. When the crude oil meets acceptable limits of purity, for example, no more than one-half of one percent BS and W, it is then ready for sale. For accounting purposes, the sale is usually deemed to take place when the crude oil leaves the operating lease.

SURFACE TREATMENT OF NATURAL GAS

Wells which are primarily natural gas wells, may or may not have associated crude oil with them. However, they generally will have variable amounts of an element called *condensate*. Condensate is very similar to high quality crude oil and normally exists in the formation under high pressure and temperature as a vapor, but as the gas reaches the surface it condenses into a liquid. This liquid is recovered, treated, and sold as a form of crude oil. The gas still contains other products which must be separated from the natural gas. This is accomplished in a large facility called a *gas plant*. A gas plant is expensive, and is usually built in an area to serve a large number of leases.

Generally speaking, any liquid which is recovered from the natural gas before it enters the gas plant is referred to as condensate. Other liquids which are removed in the gas plant, for example, butane, propane, and natural gasoline, are referred to as *natural gas liquids*. These NGL's are recovered and sold as is the purified methane. The distinction between condensate and natural gas liquids is crucial for accounting purposes because different tax regulations apply.

SUMMARY

The oil and gas industry is large, costly and complex. In this section we have attempted to define a series of terms and identify some considerations that are important to understanding the accounting issues of later chapters. Remember that all of the concepts discussed here are highly compex, and that a thorough study will be necessary for complete understanding of any of the topics in this chapter.

Chapter 3

Flow of Information in Oil and Gas Producing Companies

In this chapter we will describe certain basic lease considerations and relationships found in most oil and gas producing companies, as well as describe both cash and information flow through the organization as a result of these relationships.

THE BASIC OIL AND GAS LEASE

Most oil and gas is produced from a property which is owned by one party and operated by another. The land owner, referred to as the *lessor*, signs a contract with the oil company, referred to as the *lessee*. The lessee explores the property and, if oil and gas are discovered, produces the product. In this section we will review the basic provisions of a typical oil and gas lease. See Appendix C for an example of a lease.

The *primary term* of the lease is usually for a stated period of time, frequently three, five or ten years, although more recently very short primary terms of leases have been signed. The primary term of the lease determines how long the lessee has to find oil and gas on the property, because the lease will be effective for the number of years specified in the primary term and as long thereafter as oil and gas are produced. If no oil and gas is found by the end of the primary end of the lease, the lease expires. That does not mean the lessor cannot sign a lease with another party or even a new lease with the original lessee, but that particular lease is no longer valid.

A *lease bonus*, or up front money, is paid to the lessor as consideration for signing the lease. This lease bonus is not an ongoing payment, but is a one-time incentive payment to the lessor that gives the lessee the right to explore the property and, if oil and gas are found, the right to produce that oil and gas.

Most leases contain a *delay rental* provision. In this provision, if the lessee does not drill within any 12 month period the company must pay a specified amount to the lessor in order to keep the lease valid. This delay rental provision is an incentive to encourage the lessee to actively explore the lease.

If a product is discovered, the lessee normally will be the party producing the oil or gas and selling it. The company then must pay a *royalty* to the lessor, which is usually stated as a portion of the total product. One-eighth has been the standard royalty, although we see leases today which have higher royalty provisions. The royalty may be taken *in cash* or *in kind*. If it is taken in cash, the lessee will sell the product and then send a check for one-eighth of the cash, subject to the provisions discussed below, to the lessor. If the royalty is taken in kind, the royalty owner takes legal title to one-eighth of all the product produced and then disposes of it at his or her own discretion. In most cases the lessee bears all of the lifting costs associated with bringing the product to the surface, treating it, and making it ready for market. The lessee also is responsible for all the costs of exploration and development on the lease. Usually no royalty is paid on any product consumed on the lease as field fuel. For example, natural gas which is burned in heater treaters to remove BS and W from crude oil would not be subject to a royalty.

These provisions have been represented as typical provisions in a typical lease. The reader must be very cautious, however, in addressing any specific situation. One of the most dominate characteristics of the oil and gas industry is that *there is no absolutely typical situation*. It is easy to find lease arrangements that go contrary to all of the provisions specified above.

BASIC TERMS

There are several basic terms we will use in the remainder of this handbook:

MINERAL INTEREST

A *mineral interest* refers to an economic interest in any mineral under the ground. If oil and gas exist in a natural formation, whoever owns any right to that oil and gas has a mineral interest; this refers to the entire interest to all of the product.

ROYALTY INTEREST

The *royalty interest* is the interest retained by the original land owner.

WORKING INTEREST

The *working interest* is that portion of the interest that belongs to the operating party of the lease. If only two parties are involved, the entire mineral interest is divided into the royalty interest which is due the land owner and the working interest which is due the oil company. The working interest may be jointly owned by several parties.

OVERRIDING ROYALTY INTEREST

Occasionally, a party who owns a working interest will sell that interest to another individual but will seek to retain part of the interest himself. This type of arrangement, referred to as an *overriding royalty interest*, simply creates an additional royalty interest out of the working interest of the lease. It usually does not affect the original royalty interest. For example, consider a lease which has a one-eighth royalty interest and seven-eighths working interest. The working interest owner sells his interest to a third party, but asks for an overriding royalty interest of one-sixteenth of total mineral interest. We then have a situation as depicted in Figure 3-1. In this situation the new working interest owner will bear all the costs of exploration and production, and there will be two royalty interests involved in this mineral interest. Overriding royalty interests are frequently used in carried interests as described below.

OTHER RELATIONSHIPS

There are several other frequent kinds of relationships between oil and gas producing companies which need to be discussed.

PRODUCTION PAYMENTS

Frequently a company that owns a working interest in a producing property will borrow funds from another entity. The repayment of those funds will come out of production of that property. This payment will usually be specified as a certain percent of the net proceeds of the working interest. Interest will accrue and payment will take place as long as the property is producing, until such time as the total loan plus interest have been repaid.

Production payments can be payments defined in terms of cash or in terms of *in-kind* payment. A production payment in cash is one where payments are made until a specific amount of princi-

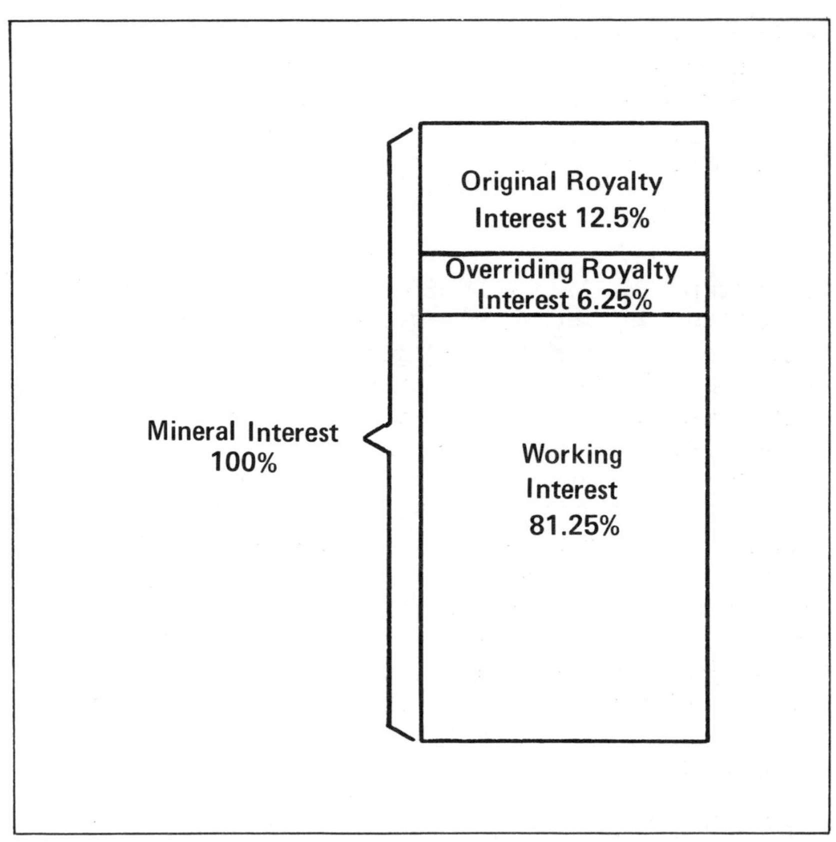

OVERRIDING ROYALTY INTEREST

Figure 3–1

ple, for example, one million dollars, plus any accrued interest at the stated rate have been repaid. If the production payment is stated in-kind, it will state that payment must be made until revenue from a certain number of barrels has been repaid. In this case the loan is not expressed in dollars but in number of barrels. Refer to ASR 257 in Appendix A for a careful discussion of the accounting treatment of production payments in-kind and production payment in cash. Simply stated, production payments in cash are treated as a borrowing, while production payments in-kind are treated as a sale of the product.

Production payments may also be made *with* or *without recourse*. If the production payment is made with recourse and the properties designated for repayment production do not produce enough to liquidate the loan, the lending party may recover the balance from the borrowing party's other assets. If, however, the production payment is made without recourse, and the property does not produce enough to liquidate the loan, the lending party has no claim on the other assets of the borrowing party and the rest of the loan is effectively forgiven. Obviously, the interest charged for a production payment without recourse is usually higher that for a production payment with recourse.

FARM OUTS

Frequently a company will have an unproved property working interest, but is unable or unwilling to expend the funds necessary to drill that property to see if proved oil and gas reserves exist. A common method of exploring that property is to *farm out* a portion of the working interest to another party who agrees to drill the well. The typical exchange works something like this: the party with the working interest, the *assignor*, will assign or transfer over to the other party, the *assignee*, a certain agreed upon portion of the working interest. In return, the assignee will drill a well on the property at his own cost. If the well turns out to be a dry hole and the property is impaired, then both parties own their agreed-upon percentages of the dry hole and everyone walks away. If, on the other hand, the assignee drills the well and finds oil and gas, the two parties have formed a partnership where each owns a certain percent of the working interest. This is then treated as a joint interest operation which will be discussed below.

For certain tax reasons it is quite beneficial for the assignee in a farm out to receive 100 percent of the proceeds until such time as he has recovered all of the costs which he spent on the original well. This point of time is called *payout*. Therefore, most farm outs today take the form of a *carried interest*, in which case the assignee or the drilling party receives 100 percent of all of the revenues and, consequently, bears 100 percent of all the lifting costs until his net proceeds from the lease equal the amount that he has invested in that lease or until payout. This form has become so popular and has become such a strong leverage in the bargaining between the assignor and the assignee, that many times today we see a carried interest property where the assignee receives 100 percent of the proceeds until he has recovered 200 or sometimes 300 percent of his cost.

JOINT INTEREST OPERATIONS

Many times two or more oil and gas producing companies will enter a lease agreement as co-owners to share the risk. If oil and gas are found on those properties, one of the owners is designated as operator. That party will develop the property, produce it, sell the product, and remit an agreed-upon portion of the proceeds to his joint owner(s). This is referred to as joint interest operations. Every joint interest property has an operator, the one who is actually physically operating the property, and one or more non-operators, parties who have a valid right to the reserves but do not participate in the actual operation of the lease.

UNITIZATIONS

Consider the situation illustrated in Figure 3-2. There are three adjacent leases owned by oil companies A, B and C. An oil and gas formation that covers all three leases is discovered by company B. The complication occurs because the reservoir extends onto the leases of other legal entities or other working interests. Each company could operate its own lease, drilling its own development wells, installing its own surface equipment, and operating autonomously from the other two companies. This would have two undesirable side effects. First, there would be significant duplicate effort in the installation of facilities. Second, each of the parties would try to drill as many wells as possible along the border lines with adjacent leases. This would be done in an effort to produce as many of the reserves as possible which are currently underground on other leases. This type of drilling war generally proves to be very inefficient and the total reserves recovered from the formation are less than if the parties cooperated with each other.

In order to overcome both undesirable affects the three parties normally agree to enter a *unitization*. After periods of negotiation and obtaining as much engineering data as possible they agree as to the total percent of recoverable reserves that exist on each property. In this example, let us assume that property A contains 30 percent of the reserves, Property B 40 percent, and property C 30 percent. Usually, but not always, the owner who has the largest percent of reserves is designated to be the operator. In this

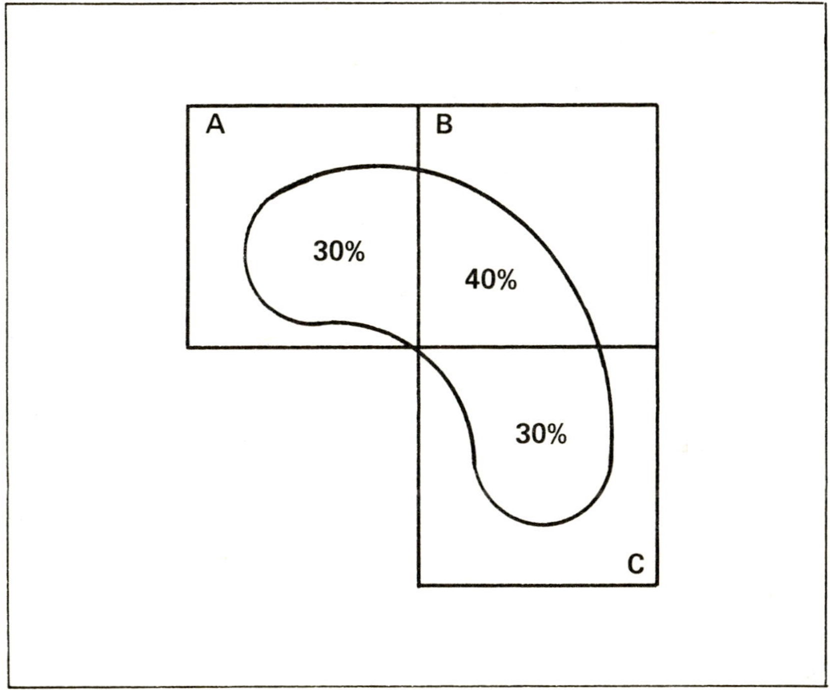

Figure 3–2

case that would be company B, who will then develop and produce the entire reservoir as if it were one property. B will produce the product, sell it, and then send 30 percent of the proceeds to A and 30 percent to C. The other members of the unit may also take their product in-kind, although they frequently take it in cash.

It should be clear that a unitization of this type is another form of joint interest operation. The primary difference here, of course, is that each party owns 100 percent of the working interest on his lease, whereas in a true joint interest operation, each party owns a specified percentage less than 100 percent of the working interest on the lease. The unitization is a very common device to arrange for a more efficient operation in the oil field.

A distinct difference should be made here between test well contributions and joint interest operations. In a *test well contribution* situation two parties own adjacent working interests. One agrees to contribute an agreed-upon share of the cost of the other party's test well cost. Seismic studies would indicate the two properties share a potentially producing reservoir. By sharing the cost of the well, the contributing party shares in the risk. If the well comes in productive, then both parties know they have reserves; if the well comes in a dry hole then both parties have participated in the cost of that particular exploratory well. The difference is, the test well contributions do not provide any of the working interest in the adjacent lease to the contributing party. The only thing that is transferred in a test well contribution is information; no legal title to working interest changes hands.

FLOW OF INFORMATION AND CASH

In most cases the flow of cash in and out of an oil company is preceeded by information that identifies how much cash is going to flow. The information may flow along with the cash. The model developed in Figure 3-3 shows the two primary in-flows and the seven primary out-flows of both cash and information. These are discussed below.

CASH IN-FLOWS

Let us assume the Black Gold Oil Company owns working interests in producing properties where it is the operator. Property proceeds are equal to the net proceeds from the sale of the product produced. These producing properties will include those where Black Gold has a 100 percent working interest and those where Black Gold has a joint working interest with another party, but is the operator. A total of 100 percent of the proceeds of these producing activities flows into the Black Gold Company. Sometimes the cash will be distributed to the various parties (royalty owner, operator, and non-operators) by the purchaser of the product.

The other primary source of cash and information in-flow comes from those properties where Black Gold is a partial owner of the working interest, but a non-operator. In these cases the operator will send Black Gold its share of proceeds. Normally this share of proceeds will be exclusive of windfall profit taxes, severance taxes and lifting costs. The operator of that property will normally withhold those items and remit only the balance to Black Gold.

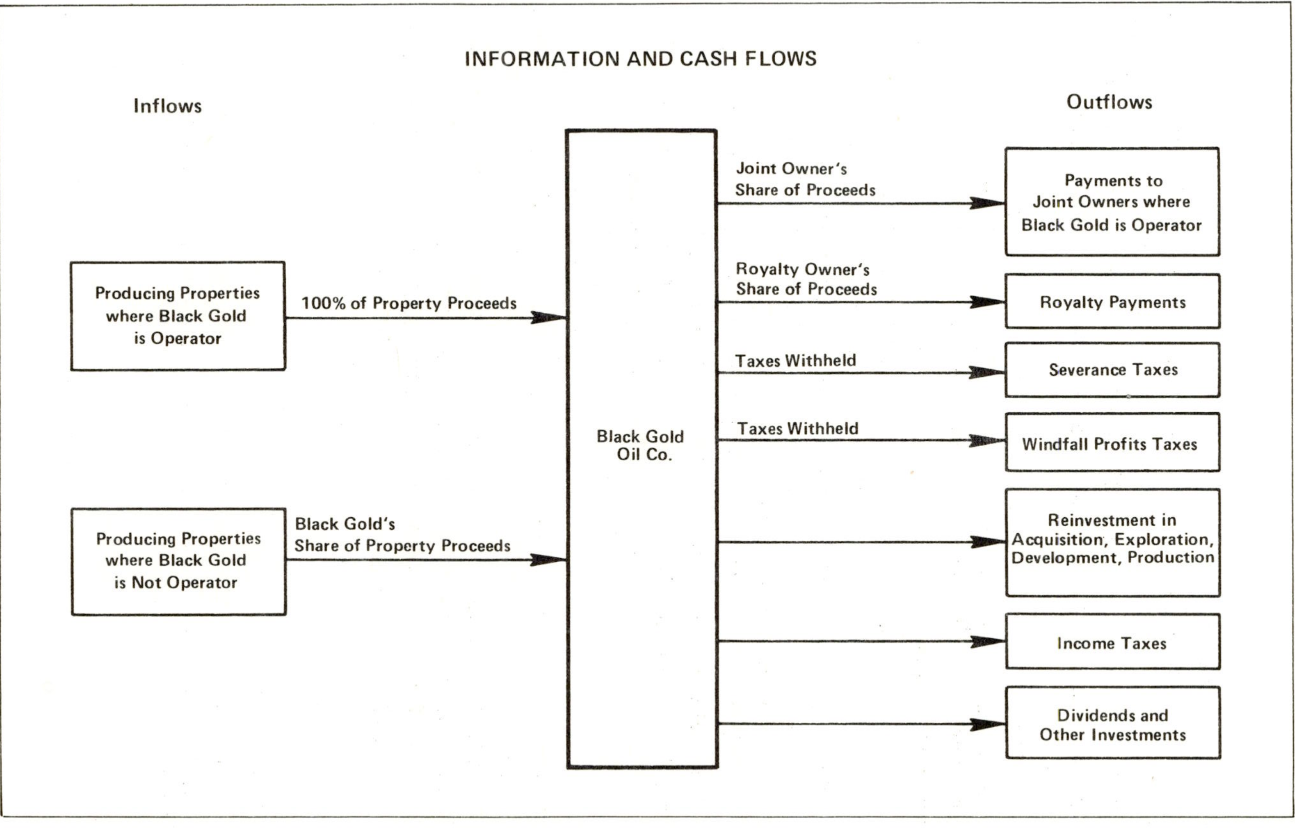
INFORMATION AND CASH FLOWS
Inflows
Outflows
Producing Properties where Black Gold is Operator
100% of Property Proceeds
Producing Properties where Black Gold is Not Operator
Black Gold's Share of Property Proceeds
Black Gold Oil Co.
Joint Owner's Share of Proceeds
Payments to Joint Owners where Black Gold is Operator
Royalty Owner's Share of Proceeds
Royalty Payments
Taxes Withheld
Severance Taxes
Taxes Withheld
Windfall Profits Taxes
Reinvestment in Acquisition, Exploration, Development, Production
Income Taxes
Dividends and Other Investments

Figure 3—3

CASH OUTFLOWS

On the other side of the coin, Black Gold must then send the proceeds to joint interest owners where Black Gold is the operator. In this case the joint owners' share of the proceeds will normally be net of severance taxes, windfall profit taxes and lifting costs.

Payments must also be made to royalty owners including overriding royalty owners. Royalty owners ordinarily do not participate in lifting costs, but these payments are normally net of severance taxes and windfall profit taxes.

Black Gold Oil Company normally will withhold *severance taxes* for all economic interests on all properties which it operates. The total amount of severance taxes will then be sent to the state along with an accounting of those taxes to each of the other economic interests involved.

Windfall profit taxes are sometimes withheld by the first purchaser and sometimes withheld by the operator. If the operator is responsible in a particular situation for collecting the windfall profit tax, then those taxes must be sent to the Federal Treasury.

Black Gold will use part of its proceeds to reinvest in acquisition of other working interests, both proved and unproved, for exploration and development activities, and as working capital for production activities.

Black Gold is also responsible for payment of any state and federal income taxes that relate to its operations.

The final cash payment refers to anything which is left over. This can be paid to the stockholders in the form of dividends, and any of that amount not paid to the stockholders can be reinvested in other activities in the company.

SUMMARY

The remaining portions of this Handbook will describe in detail the specific cash flows and the accounting treatment for them.

Chapter 4

Full Costing Versus Successful Efforts

In this chapter we will review the history of the two primary methods of accounting by oil and gas producing companies, *full costing*[1] and *successful efforts*, discuss the primary differences between the two methods, and illustrate those differences with a comprehensive example.

This difference between methods is more than a subtle one. Let it suffice at this point to say that the same company using the same actual production results would show dramatically different net incomes and asset levels on the balance sheet if the company used the two different methods of accounting.

We will see later that both methods are used throughout all sizes of companies in the oil and gas industry. Since our industry is so totally dominated by joint interest operations, anyone who works with oil and gas accounting will certainly be involved with projects that involve both full costing and successful efforts companies. Therefore, it is important that those involved in oil and gas accounting be familiar with both methods regardless of which method is used by their own company.

HISTORICAL BACKGROUND

Prior to 1960, virtually all companies followed a form of accounting which is today known as successful efforts. Briefly stated, in successful efforts accounting only costs associated with specific proved reserves are capitalized. All other costs are expensed in the year incurred. Refer to Chapter One for a discussion of capitalization versus expensing of cost. There were many forms of successful efforts accounting, and none of the accounting bodies had issued any pronouncements to give guidance on which of these forms are preferable. However, the preponderance of accounting practice was basically successful efforts in nature.

[1]The term full costing also is referred to frequently as full cost. Both forms are widely used and both are acceptable.

In the late 1950's a new method of accounting was developed and quickly adopted by a large number of companies. This method is referred to as full costing accounting. Briefly stated, companies that follow full costing capitalize all costs associated with property acquisition, exploration and development, including those costs which are not associated with proved reserves such as the cost of a dry hole exploratory well.

The example in this chapter will clearly illustrate the very wide degree of difference between these two methods of accounting.

As a generalization, we could say that a company that is engaged in an aggressive program of acquisition and exploration will show a higher net income and also a higher level of assets under full costing than under successful efforts. For this reason the full costing method immediately became popular among young, aggressive companies. The method soon spread and was widely practiced throughout the industry. Table 4-1 shows clearly that both methods of accounting are today followed by companies of all sizes throughout the industry, and that the division total is nearly equal.

One of the basic principles of accounting is that of comparability; that is, if an investor wishes to evaluate several different companies in the same industry, he or she should be able to rely on financial statements of those companies being prepared under the same accounting rules. Unfortunately, this level of comparability simply does not exist in the oil and gas industry because of the wide difference in accounting rules for full costing and successful efforts. For that reason, the Accounting Principles Board looked at the problem during the mid-1960's and was unable to decide that one method was clearly better than the other. The Board took no official action. When the Financial Accounting Standard Board was created in the early 1970's oil and gas accounting was not put on the initial agenda.

Congress interfered in 1975. The Energy Policy and Conservation Act of that year was a congressional response to the shortages caused by the 1973 Arab oil embargo. In that Act, Congress asked the S.E.C. to develop accounting rules for all persons engaged in the production of crude oil and natural gas. These rules were ultimately to have formed the accounting basis for a national energy data base which would include certain financial information.

COMPANIES THAT HAVE OIL AND GAS PRODUCING OPERATIONS
AND FILE ANNUAL REPORTS WITH THE SEC

COMPARISON OF ACCOUNTING METHODS
SHOWING TOTAL REVENUES OF U.S.COMPANIES

	NUMBER OF COMPANIES			PERCENT	
	Successful Efforts	Full Cost	TOTAL	Successful Efforts	Full Cost
U.S. COMPANIES					
Total Revenues					
Over $1 Billion	35	28	63	56%	44%
$100-999 Million	32	34	66	48%	52%
$10-99 Million	23	34	57	40%	60%
Under $10 Million	43	28	71	61%	39%
Total U.S. Companies	133	124	257	52%	48%
CANADIAN COMPANIES	9	26	35	26%	74%
TOTAL	142	150	292	49%	51%

Source: Financial Accounting Standards Board, "Additional Comments of the FASB to the Securities and Exchange Commission; Accounting Practices—Oil and Gas Producers, SEC File S7-715" May 31, 1978.

Table 4—1

After an interminably long series of hearings, proposed rules, supplementary disclosures, and other actions we now have a number of rules in effect.[2] Rules for companies following the successful efforts method of accounting are contained in the Financial Accounting Standards Boards document, Statement of Financial Accounting Standards Number 19, "Financial Accounting and Reporting by Oil and Gas Producing Companies".[3] Sucessful efforts companies that are required to file statements with the S.E.C. must follow ASR No. 257, which is essentially a verbatim reproduction of the successful efforts rules contained in Statement of Financial Accounting Standards No. 19. Full costing companies that are S.E.C. registrants must follow the rules established in ASR No. 258. Full costing companies that are not S.E.C. registrants have no specific guidelines and are thus able to follow essentially any set of rules they wish. The complete text of both ASR's No. 257 and 258 are contained in Appendix A of this text.

It is probable the profession will someday adopt a single method of accounting for oil and gas producing companies. That method may continue to be one of the historical cost methods, such as successful efforts or full costing, or more probably, it will be a method which is based not on the historical cost of acquiring reserves, but rather on some value of those reserves. An experimental form of this value based accounting was tried by the S.E.C. The method called *Reserve Recognition Accounting* (RRA) was not adopted as the primary financial statement method because of the inability of reservoir engineers to precisely and accurately measure reserve quantities. However, this method of measuring results of operations is still required as supplemental

[2]For a complete discussion of the rules that had taken place see Robert J. Koester and Harold M. Nix "Oil and Gas Accounting; a Discipline in Transition," *CPA 80*, September 1980, pp 26-28.

[3]The effective date of this document has been indefinately suspended by Statement of Financial Accounting Standards Number 25. Therefore, companies are not strictly required to follow this document. However, the successful efforts rules in this document are becoming the generally accepted rule for successful efforts companies, and a successful effort company which does not follow these rules yet wishes to have audited financial statements should be prepared to justify a deviation from the rules.

disclosure outside of the financial statements by all S.E.C. registrants.[4]

SPECIFIC DIFFERENCES BETWEEN THE METHODS

It is generally held that the difference between the two methods revolves around the treatment of costs for exploratory dry holes: successful efforts companies expense those costs and full costing companies capitalize them. This is a difference between the two methods, but it is an over-simplification. Unfortunately, this over-simplification is very widely held.

Consider, for example, this quote from the appendix to the Energy Information Administration's report on financial performances of oil and gas producing companies of 1978.

> At this time, petroleum producing companies reporting to the Securities and Exchange Commission, which include all FRS reporting companies, are permitted to choose between two accounting methods—*full cost* and *successful efforts.* The two methods differ from one another in the treatment of the cost of dry exploratory holes.
>
> Under full cost, the cost of a dry exploratory hole is capitalized and then amortized to the income statement over the production life of successful wells, which is to say it is spread over many future years. Thus the capitalized costs of both dry and successful wells are reflected in the balance sheet as part of producing properties.
>
> Under successful efforts, the cost of a dry exploratory hole is written off to expense in the year the drilling is determined to be unsuccessful. There is then no capitalized cost of such dry exploratory holes carried on the balance sheet.

This view of the difference between the two methods is widely held but does not examine the real differences. We can see that there are two primary areas where the two methods differ. The first is in the treatment of capital expenditures. It is true that suc-

[4]For a complete discussion on these issues see Robert J. Koester, Paul Munter, and Thomas A. Ratcliffe, "Reserve Recognition Accounting Made Simple," *Oil and Gas Tax Quarterly*, Fall 1981.

cessful efforts companies do expense most of those costs, while full costing companies capitalize them and then spread the costs over production of whatever reserves are found. An even more important difference may be the size of the cost center that is used to calculate the amortization of those costs. We will examine these two differences in detail.

THE CAPITALIZATION DECISION

There are four classifications of costs in oil and gas production:

1. *Acquisition Costs* are all those costs incurred in order to acquire legal title to the working interest in the mineral property. These costs are essentially lease bonuses, or if an existing working interest is purchased, the lease purchase cost.
2. *Exploration Costs* are all of those costs incurred to resolve doubt as to whether or not proved reserves actually exist on the property. These costs include geological and geophysical costs as well as exploratory drilling costs.
3. *Development Costs* are all those costs incurred after proved reserves are determined to exist on the property, up to the point where the property is capable of producing reserves. These costs include completing the discovery well (such as casing, cementing or perforating) and the cost of development wells plus all surface equipment that is installed.
4. *Production Costs* include all those costs incurred once the property is capable of producing until the oil or gas leaves the lease. These costs are primarily lifting costs, treatment costs, and the two primary taxes, severance taxes and crude oil windfall profit tax.

Refer to Figure 4-1. Notice that in full costing accounting all acquisition, exploration and development costs are capitalized within the cost center. That is, these costs are declared to be assets as soon as they are incurred, and then when any production does take place, they are charged off or expensed against revenue on a per barrel or per unit basis. All production costs for full costing companies are expensed as incurred.

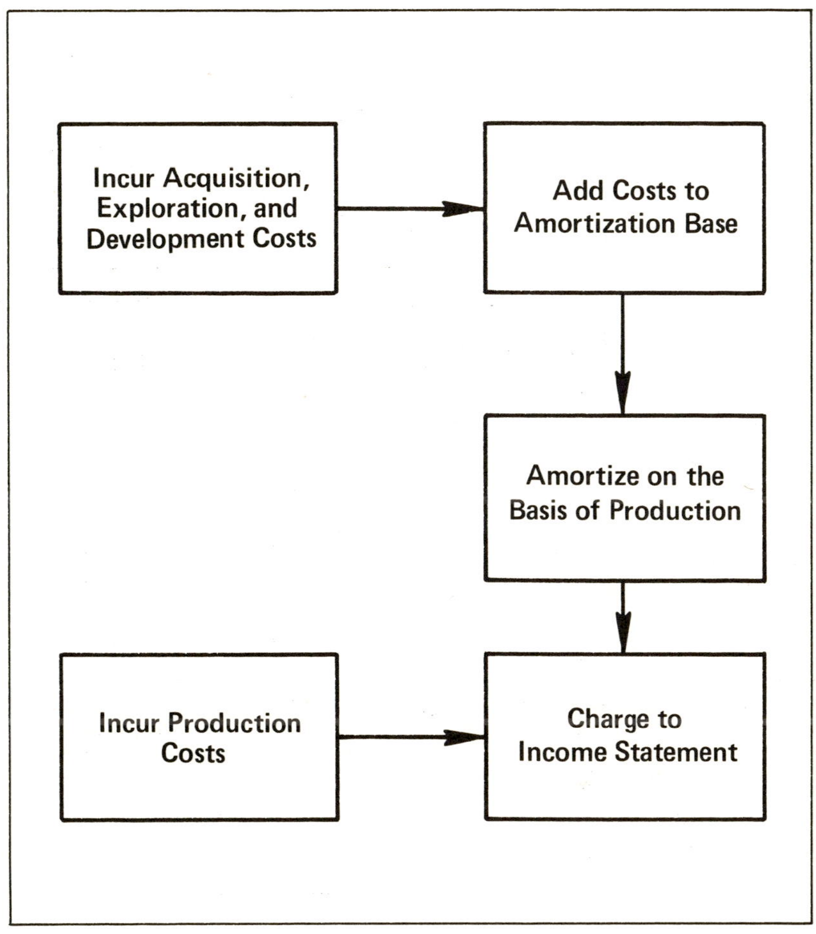

FULL COSTING OVERVIEW
Figure 4—1

Now compare this treatment with Figure 4-2. Here we see that for a successful efforts company, production costs also are charged to expense as incurred and development costs also are capitalized. The major difference occurs for acquisition and exploration costs. In each case the costs are held in a suspense status until doubt is resolved as to whether or not proved reserves exist on the property. If it is determined that proved reserves do ex-

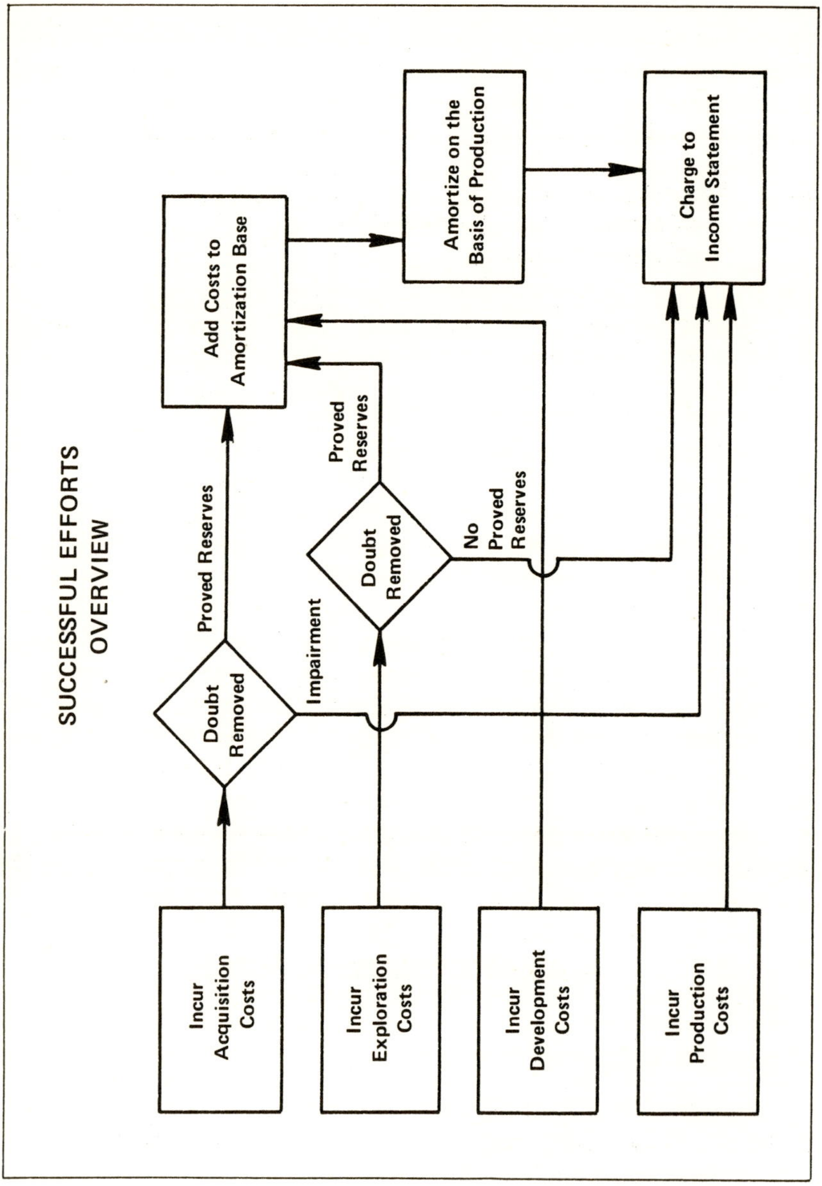

SUCCESSFUL EFFORTS
OVERVIEW
Incur Acquisition Costs
Incur Exploration Costs
Incur Development Costs
Incur Production Costs
Doubt Removed
Proved Reserves
Impairment
Doubt Removed
Proved Reserves
No Proved Reserves
Add Costs to Amortization Base
Amortize on the Basis of Production
Charge to Income Statement

Figure 4–2

ist on the property, the costs are added to the amoritization base. That is, they are declared to be an asset and the cost of that asset is amortized on a per barrel basis. However, if the property is *impaired*,[5] the acquisition costs are written off as an expense.

Any exploration costs which do not find specific new reserves (this would include G and G costs, delay rentals, test well contributions and all exploratory dry holes) are written off as an expense. The only exploration cost which is capitalized under successful efforts is the exploratory well that finds new reserves.

These two flow charts illustrate the primary difference in the treatment of capitalized cost for the two methods.

THE COST CENTER CONCEPT

However, perhaps even more important than the capitalization decision is the decision of the size of the cost center. This concept is a little more challenging so we will illustrate it with an example. Assume we have three properties as shown in Figure 4-3, A, B and C. We will further suppose that there is a small formation that covers both properties A and B and contains approximately 1,000,000 barrels of proved reserves on property A and 2,000,000 barrels of proved reserves on property B. Property C contains no proved reserves. Let's further assume that we have capitalized costs of \$3,000,000 on property A, \$4,000,000 on property B, and that we have spent \$2,000,000 on property C. We now have three options or alternatives for determining the cost center size.

Assume under the first alternative that we establish as our cost center size each individual property. We must then calculate an amortization rate. The rate is equal to the fraction:

$$\text{rate} = \frac{\text{dollars of capitalized cost}}{\text{barrels of proved reserves}}$$

Property A would then have a rate of

$$\frac{\$3{,}000{,}000}{1{,}000{,}000 \text{ bbl}} = \$3 \text{ per bbl}$$

[5]See Chapter Five for a complete discussion of the criteria for impairment.

Property B would have the rate of

$$\frac{\$4{,}000{,}000}{2{,}000{,}000\text{ bbl}} = \$2\text{ per bbl}$$

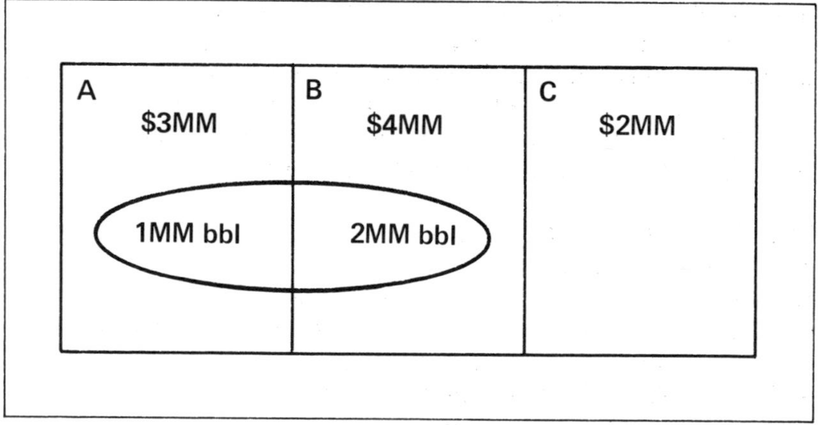

Figure 4–3

Property C has no proved reserves; therefore, if it is a separate cost center and followed the accounting convention of lower of cost or market, we would have no amortization rate at all. Property C would have no asset value. Notice that this alternative shows a different rate for the different properties based on a comparison of the amount spent versus the amount discovered.

In the second alternative, let us assume each cost center is defined as all properties on which there are proved reserves. One cost center would consist of both properties A and B, and property C would be a separate cost center. The rate for properties A and B would then be

$$\frac{\$7{,}000{,}000}{3{,}000{,}000\text{ bbl}} = \$2.33\text{ per bbl}$$

Property C would still have no amortization rate for the previous reasons cited. Notice this simply is an averaging process of the amortization rates of those properties which do have proved reserves.

In the third alternative, we set our cost center equal to all properties whether they have proved reserves or not. In this example, we have only one cost center consisting of properties A, B, and C. The costs now have been increased so that the rate is

$$\frac{\$9,000,000}{3,000,000 \text{ bbl}} = \$3 \text{ per bbl}$$

Notice that this averages in the cost of the non-productive property with the productive properties.

Then we can see that as the cost center becomes larger and includes more properties, the averaging process becomes more accelerated. Successful efforts companies use one of three cost centers: the individual property or lease, the individual reservoir, or a field which is a collection of reservoirs. These are relatively small cost centers. Consequently, for most successful efforts companies, properties that do not result in proved reserves are written off to an expense because of the lower of cost or market rule.

By contrast, full costing companies following ASR No. 258 must use as their cost center the entire United States. Other cost centers for companies which have international operations can be each individual country, or a group of countries if they can be logically grouped.

As we will see in the example, this cost center decision may have an even more profound affect on net income than the capitalization decision.

For example, refer to Figure 4-4. Assume in this example that we have acquired four leases during year one. These leases are the Acme, the Bledsoe, the Carter and the Discovery leases. Before we signed any leases we conducted preliminary seismic studies on a large number of acres. These are referred to as preacquisition costs and are not tied to any of the four specific leases we signed. The acquisition costs refer to the lease bonus paid for each of the four leases. The geological and geophysical costs are additional seismic studies conducted after the leases were signed. These results showed positive formations on Bledsoe, Carter, and Discovery. It should be noted here that the G and G efforts cannot find reserves but only rock formations which may contain reserves. The G and G results on the Acme lease did not disclose any formation which would indicate that we should drill.

FC AND SE EXAMPLE

	LEASES			
	Acme	Bledsoe	Carter	Discovery
YEAR 1				
Pre-acquisition Costs $50,000				
Acquisition Costs	$ 6,000	$ 8,000	$20,000	$ 2,000
G and G Costs	20,000	15,000	25,000	10,000
G and G Results	negative	positive	positive	positive
Exploratory Wells	–0–	2@$800,000 ea.	3@ $1,200,000 ea.	1@$600,000
Drilling Results	none	dry holes	dry holes	discovery
WI Reserves	–0–	–0–	–0–	2MM bbl
YEAR 2				
Exploratory Wells	–0–	–0–	1@$1,400,000	
Drilling Results			dry hole	
Development Costs	–0–	–0–	–0–	$3,000,000
Production (WI)				120M bbl
Lifting Costs				$ 960,000
Sales Price				$40/bbl

Figure 4–4

Two wells were drilled on the Bledsoe property, three on the Carter, and one on the Discovery lease. All the wells were dry holes except the one on the Discovery lease. Two million barrels of proved reserves on that lease were discovered. These barrels are referred to as working interest barrels, in that this example does not consider the effects of royalty interest.[6] During the second year of this example an additional exploratory well was drilled on the Carter lease and was also a dry hole.

Development costs of $3,000,000 were spent to bring the Discovery lease to a point where it could be produced. These consisted of development wells as well as surface equipment; 120,000 barrels of production occurred during year two. Other relevant data are shown in Figure 4-4.

The Acme lease was impaired as soon as the G and G results showed no formation which was likely to contain oil and gas. Significant doubts still remain about the Bledsoe lease, so that it has not been impaired. After the fourth dry hole was drilled on the Carter lease, it was deemed to be impaired. Doubt was resolved on the Discovery lease as soon as the exploratory well was completed and it was determined that proved reserves exist on that property. Ordinarily a delay rental would need to be paid on the Bledsoe lease to keep it active. We'll deal with delay rentals in Chapter Six.

SUCCESSFUL EFFORTS—YEAR ONE

Refer to Figure 4-5. There was no revenue during this year, but there were several items which should be expensed for successful efforts purposes. The pre-acquisition costs only indicated which properties might have potential reserves and are expensed. The same is true for G and G costs. The lease bonus paid for Acme is considered to be an expense this year because that property has been impaired. Again, remember we must follow the lower of cost or market rule. Since Acme is impaired, it is deemed to have no market value; therefore, all of this cost center must be written off in the year of impairment.

All of the dry hole costs on the exploratory wells are expensed in the year in which those wells are determined to be dry holes.

[6]See chapter Seven for a complete discussion of accounting for royalty interests.

Thus, the net income before taxes for successful efforts in year one is a very large loss. This will be true in any year in which a successful efforts company has a large number of exploratory wells which are generally unsuccessful.

ABC OIL COMPANY
INCOME STATEMENT (SE)
YEAR 1

Revenue		–0–
Expenses:		
Pre-acquisition	$ 50,000	
G and G	70,000	
Impairment (A)	6,000	
Dry Holes	5,200,000	$ 5,326,000
Net Income Before Taxes		$(5,326,000)
Plant, Property, and Equipment:		
Unproved Properties (B & C)	$ 28,000	
Proved Properties (D)	2,000	
Wells, Equipment, Etc. (D)	600,000	
Total Plant, Property, and Equipment	$ 630,000	

Figure 4–5

The successful efforts company would show a relatively modest value for the assets on the balance sheet. Both B and C are in doubt at the end of the year and so their acquisition costs are included as assets under unproved properties. The acquisition cost for property D also is included, but it is classified as a proved property since it has been determined to have proved reserves. The discovery well on lease D is now considered to be an asset which is carried forward on the balance sheet, and it is shown under the classification of wells and related equipment and facilities.

FULL COSTING — YEAR ONE

Refer to Figure 4-6. Since all of the costs incurred during year one were in the pursuit of proved reserves, they are all capitalized for full costing purposes. We can treat all four properties as one cost center; therefore, the market value of the reserves that we did discover on lease D is adequate to cover all of the capitalized cost. This makes a very simple income statement in that there were no revenues or expenses.

ABC OIL COMPANY
INCOME STATEMENT (FC)
YEAR 1

Revenue	–0–
Expenses	–0–
Net Income Before Taxes	–0–
Plant, Property, and Equipment:	
Pre-acquisition Costs	$ 50,000
Acquisition	36,000
G and G Costs	70,000
Wells, Equipment, Etc.	5,800,000
Total Plant, Property, and Equipment	$5,956,000

Figure 4–6

All the costs are shown on the balance sheet. They were necessary (under full costing philosophy) in order to find the proved reserves that were ultimately discovered on the Discovery lease; therefore, we can see a substantially higher asset value for the full costing company than we have for the successful efforts company.

SUCCESSFUL EFFORTS — YEAR TWO

See Figure 4-7. There was a modest amount of production this year so we do show some revenue; however, we have impaired the Carter lease and so its acquisition cost is expensed. Notice the timing here. The lease was acquired in year one, and doubt still remained at the end of that year so we did not expense it until year two when we resolved that doubt. Had doubt been resolved in favor of proved reserves the $20,000 would have been reclassified as another asset, proved properties. However, in this case since the doubt was resolved in favor of impairment the $20,000 is considered to be an expense and is charged against income in that year. The additional dry hole is shown as an expense as are all of the production or lifting costs.[7] It is necessary to charge as an expense a portion of the costs which are capitalized within the cost center that produces reserves; therefore, we must consider which costs have been capitalized in the Discovery lease. The acquisition cost, the discovery well, and the development cost all fall in this category. Therefore, we must take these costs and divide them by the total number of barrels of reserves discovered. This comes to $1.80 per barrel for the amortization rate. We reduce income by an amount equal to the production times that rate.[8] Notice that in the second year when we did have some revenue the income picture looks much better than it did in year one.

There are several changes now on the balance sheet for the successful efforts company. Since we have written off the acquisition costs of the Carter lease, only the Bledsoe lease remains in the category of unproved properties. The original acquisition cost of the Discovery lease still remains under proved properties, although it has been amortized. Wells, related equipment and facilities include all of the capitalized costs on the Discovery lease except for the acquisition cost. The amortization number that is subtracted from the capitalized amount includes amortization of both acquisition and other capitalized costs.

[7]Remember these lifting costs also include working interest share of severance taxes and crude oil windfall profit taxes.

[8]One of the most pervasive characteristics in oil and gas production is that the number of barrels of reserves discovered is only an estimate that is frequently substantially revised. We will deal with the revision of reserve estimates in detail in Chapter Eight.

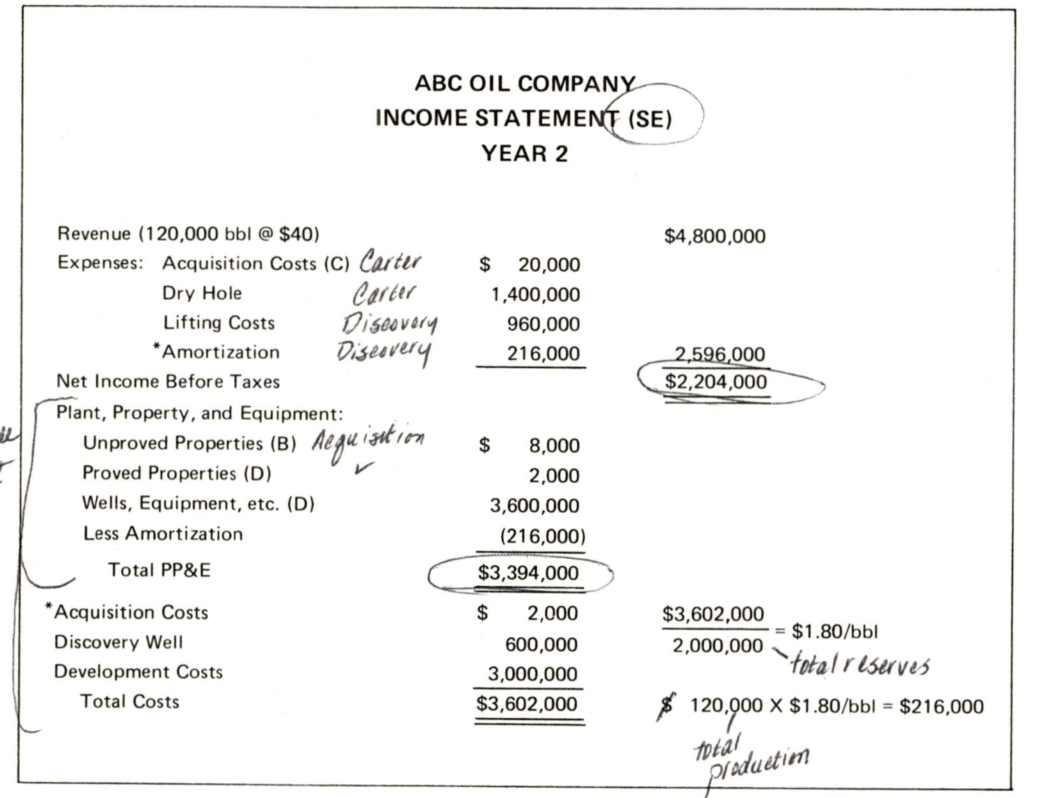

ABC OIL COMPANY
INCOME STATEMENT (SE)
YEAR 2

Revenue (120,000 bbl @ $40)			$4,800,000
Expenses:	Acquisition Costs (C)	$ 20,000	
	Dry Hole	1,400,000	
	Lifting Costs	960,000	
	*Amortization	216,000	2,596,000
Net Income Before Taxes			$2,204,000
Plant, Property, and Equipment:			
	Unproved Properties (B)	$ 8,000	
	Proved Properties (D)	2,000	
	Wells, Equipment, etc. (D)	3,600,000	
	Less Amortization	(216,000)	
	Total PP&E	$3,394,000	
*Acquisition Costs		$ 2,000	
Discovery Well		600,000	
Development Costs		3,000,000	
	Total Costs	$3,602,000	

$$\frac{\$3{,}602{,}000}{2{,}000{,}000} = \$1.80/\text{bbl}$$

120,000 X $1.80/bbl = $216,000

Figure 4–7

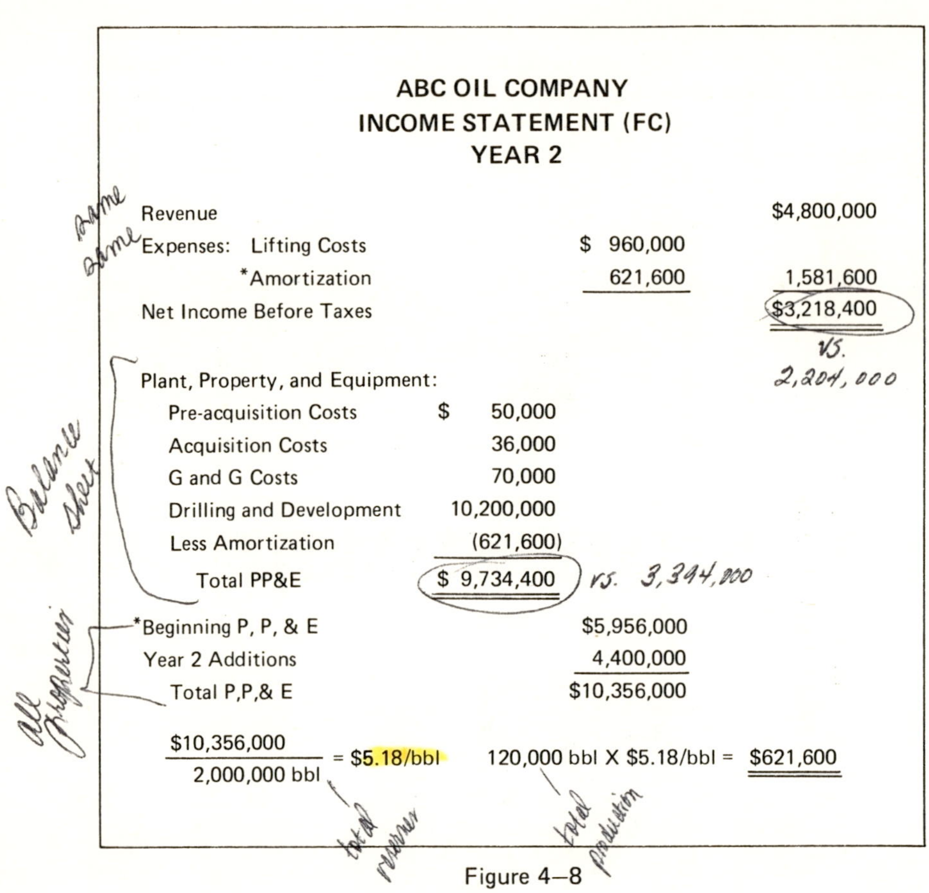

ABC OIL COMPANY
INCOME STATEMENT (FC)
YEAR 2

Revenue			$4,800,000
Expenses: Lifting Costs		$ 960,000	
*Amortization		621,600	1,581,600
Net Income Before Taxes			$3,218,400
Plant, Property, and Equipment:			
Pre-acquisition Costs	$ 50,000		
Acquisition Costs	36,000		
G and G Costs	70,000		
Drilling and Development	10,200,000		
Less Amortization	(621,600)		
Total PP&E	$ 9,734,400		
*Beginning P, P, & E		$5,956,000	
Year 2 Additions		4,400,000	
Total P,P,& E		$10,356,000	

$$\frac{\$10{,}356{,}000}{2{,}000{,}000\text{ bbl}} = \$5.18/\text{bbl} \qquad 120{,}000\text{ bbl} \times \$5.18/\text{bbl} = \$621{,}600$$

Figure 4–8

FULL COSTING — YEAR TWO

Refer to Figure 4-8. The revenue amounts for the full costing and successful efforts companies will be the same. However, we do not expense the dry hole or the acquisition cost of the impaired Carter lease. Lifting costs are the same as they would be for successful efforts. Amortization under full costing almost always is higher than it is under successful efforts because more costs are included in the capitalized cost center. Notice that the net income before taxes under full costing is higher than it is for successful efforts.

Plant property and equipment also is substantially higher. All the pre-acquisition costs, acquisition, and geological and geophysical costs remain in the capitalized cost center even though some of those properties have been impaired. All the drilling and development costs also remain within that capitalized cost center. The amortization is higher because the amortization rate is higher.

GENERAL DIFFERENCES BETWEEN FULL COSTING AND SUCCESSFUL EFFORTS

We can illustrate the differences between full costing and successful efforts by considering the life of one lease. If a company had only that lease and the lease proved to be successful, its *decline curve* or *production curve* might look like the example in Figure 4-9. This curve assumes production began in year four. Production increased for the next few years until it peaked out, and then declined at a constant rate of decline. In year 19, it is assumed that the property is no longer feasible to produce; that is, lifting costs exceed revenue, and the property is *plugged and abandoned*.

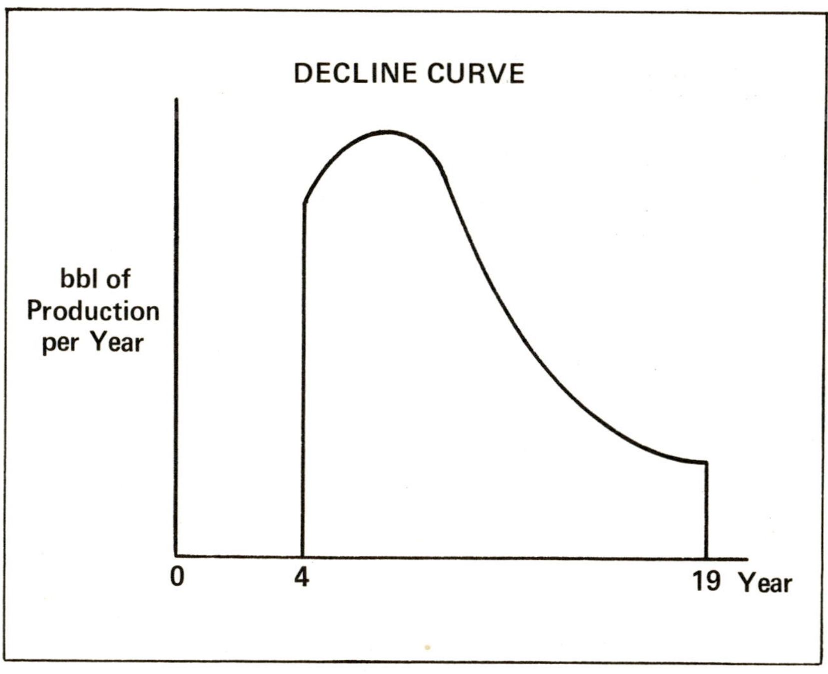

Figure 4–9

Under successful efforts accounting this lease would show very substantial losses during the first four years. These will be of variable rates depending on which years wells were determined to be dry holes, when delay rentals were paid, or when G and G costs incurred. In year four a net income would begin and it would then decline in the same shape as the production curve as in Figure 4-10.

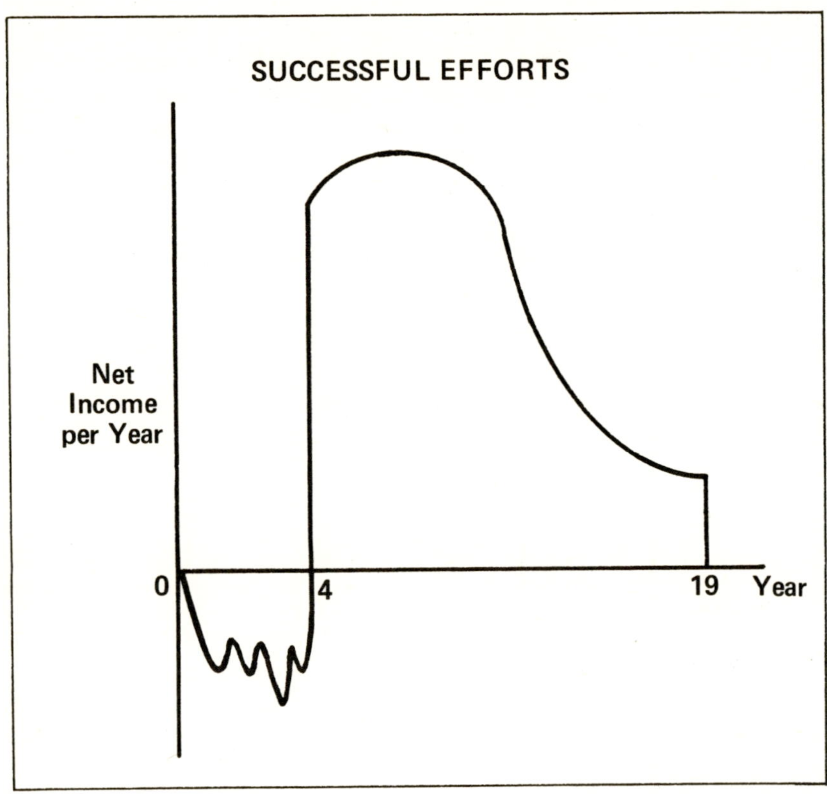

Figure 4—10

Under full costing all expenditures during the first four years would be capitalized so there would be no revenue and no expenses during those first four years, and net income would be equal to zero. In year four when production begins so would net income. The shape of that curve will be the same as for successful efforts, except that it will be lower. See Figure 4-11. The reason for this is that the amortization rate for successful efforts is lower than the amortization rate for full costing.

SUMMARY

The important point is that net income over the life of the lease will be the same under both methods of accounting. There is however, a very substantial difference in timing. Since much of our accounting data and accounting efforts deal with determining the net income for each period (quarter or annual), this matter of timing becomes crucial. As a general rule, full costing delays the recognition of most expenditures as an expense but then has a higher amortization rate and, consequently, a lower net income in later years than does successful efforts.

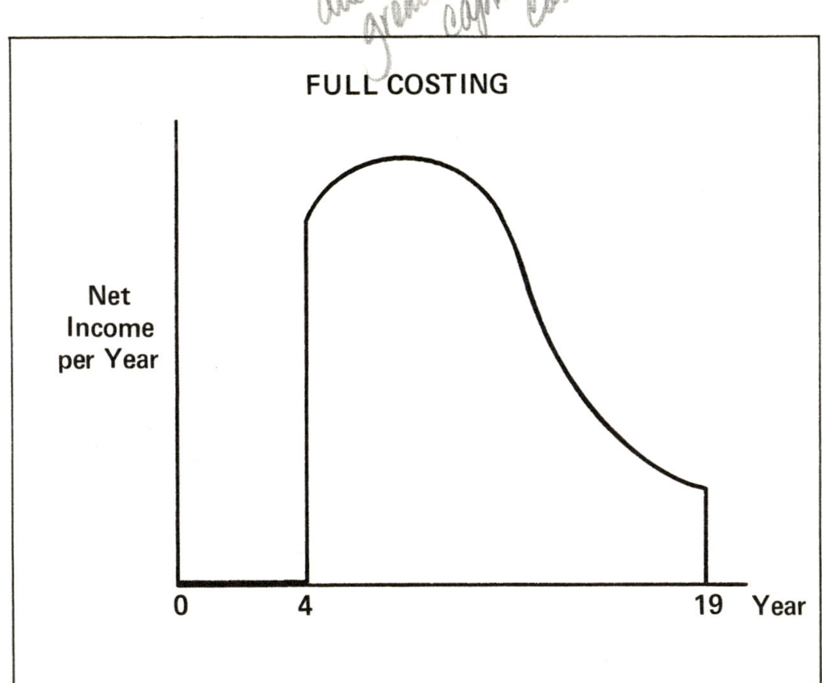

Figure 4—11

Chapter 5

Accounting for Pre-Acquisition and Acquisition Costs

In this chapter we will begin our detailed examination of the accounting treatment for each of the specific kinds of costs: first are those costs incurred by the company prior to acquisition of the lease, and then are the costs associated with acquiring the mineral interest to the oil and gas reserves. We also will discuss what happens to those costs as the property is explored.

NATURE OF PRE-ACQUISITION COSTS

Prior to signing a lease with a land owner the oil company will frequently conduct preliminary seismic studies on a property. In most cases the company will conduct these seismic studies over a very large number of acres and then attempt to sign leases only on those tracts of land which appear to have favorable formations. Two cash expenditures are important in this respect.

The first expenditure is that for *shooting rights*. This is a cash payment to the land owner for the right to come on the land and conduct the seismic studies. There is no further obligation on the part of the land owner; that is, no interest in any mineral ultimately discovered is obtained through the payment of a shooting right. Therefore, these payments are relatively small on a per acre basis.

The other major cash expenditure is for the seismic studies themselves. Frequently an oil company will purchase the seismic results from an independent source, either a consulting firm or another oil and gas company, or the company may conduct their own studies. If the seismic studies are purchased externally then the purchase cost is the cost we will treat here as the *G and G costs*.[1]

[1]Geological costs refer to those costs in identifying specific structures, their size and shape. Geophysical costs are associated with determining the properties of the rocks ultimately examined.

In a situation where a company has its own seismic department the costs we are concerned about are the actual direct operating cost of that department, plus an appropriate share of overhead which is allocated on a systematic and rational basis.

NATURE OF ACQUISITION COST

There are three forms of acquisition costs. The first is the purchase of an existing working interest in a lease which has proved reserves. Recall that a working interest is an asset which can be sold by its owner to any willing buyer. This purchase can be from the original lessee or a subsequent purchaser of the working interest. If the property has proved reserves it is considered to be a *proved property*.[2] In the case of such a purchase, the acquisition cost is the actual consideration given up to acquire that working interest.

The second type of acquisition cost is the purchase of an existing working interest, when the property does not have proved reserves. In this case we are also buying the working interest, but it is treated quite differently for accounting purposes. Naturally, if we buy a proved property we are paying for the reserves which have already been discovered and the purchase price will be significantly higher. In the purchase of an unproved property what we are buying is the potential for proved reserves.

The more common type of acquisition is that of the original working interest. In this case, the oil company is dealing directly with the original lessor. The costs which are associated with this acquisition include the lease bonus paid to the lessor plus any legal fees, attorney fees, or filing costs which are associated with finalizing the legality of the lease.

A special accounting problem exists when an oil company acquires a property *in fee*; that is, a company acquires not only ownership of all the mineral interests but also the surface rights. In this case the total purchase price paid for the property must be allocated to that portion which represents surface rights, and the remainder to the mineral interest. The costs associated with the mineral interest are either impaired or amortized over production. Costs associated with surface rights are usually not written off. The most common method of this allocation is to estimate the fair

[2]We will discuss the determination of proved reserves at length in Chapter Six.

market value of surface rights by examining sales of nearby properties. This amount is then allocated to the surface rights and the remainder is allocated to the mineral interest.

BASIC ACCOUNTING RULES

For full costing companies, all pre-acquisition and acquisition costs are capitalized in the appropriate cost center. Remember that for domestic oil and gas companies, this cost center represents the whole United States. As long as enough proved reserves are present somewhere in the cost center to justify carrying the costs as an asset, all costs associated with this property acquisition are capitalized. They are then amortized over any production that does occur in the entire cost center.

The process is somewhat more complicated for a successful efforts company. All pre-acquisition costs are immediately written off to an expense. These are actually exploration costs. The shooting rights only identify potential formations; they do not actually find new additional proved reserves. Therefore, consistent with the general rules of successful efforts accounting, these costs are written off as incurred. The costs of acquiring a property, however, are capitalized when first incurred.

If a successful efforts company acquires a mineral interest in a proved property the costs are capitalized as proved properties, and then written off over any production which occurs in the appropriate cost center.

However, if a successful efforts company acquires a mineral interest in an unproved property, whether it is the purchase of an existing working interest or the original acquisition of the lease from the original lessor, the costs are initially capitalized in an account titled *Unproved Properties*. This account is shown as an asset on the balance sheet but is not amortized over production. The costs remain in this account until one of two events occur.

If the unproved property is impaired, that is, if we become less certain that proved reserves will eventually be discovered on the property, the acquisition costs are written off to a loss account and reduce net income in the year in which impairment occurs.

On the other hand, if proved reserves are discovered on the property, the acquisition costs are transferred to a *Proved Property account*. In this account, the costs are written off over any production that occurs within the cost center; thus, it is obvious that the acquisition cost of an unproved property must be written off against income and shown as a reduction of income, either through the impairment process or through the amortization of proved reserves. As a practical matter acquisition costs can remain in Unproved Properties only as long as the primary term of the lease.

THE IMPAIRMENT PROCESS

Under certain circumstances we must consider *impairment* for full costing accounting, but those unique circumstances are not dealt with in this Handbook; however, impairment is especially important for successful efforts accounting.

We must first compare the impairment process to the *abandonment* process. The two are quite different in that impairment refers to a change in management's attitude about the potential that proved reserves will be discovered on the property. On the other hand, abandonment refers to giving up legal title to the property.

Impairment has occurred if management is less confident than it used to be that proved reserves will be discovered on the property. This can occur because of the drilling of dry holes, because of adverse geological and geophysical results, or because of adverse drilling results on nearby properties. In successful efforts accounting, all or a portion of acquisition costs of that cost center must be written off to a loss and thereby reduce this period's net income. The offsetting entry is to an allowance account that is contra to Unproved Properties; that is, it is subtracted from the balance of the Unproved Properties account.

For example, assume a successful efforts company has an unproved property cost center with an acquisition cost of $10,000 and that, because of adverse drilling results, the company is only 75 percent as sure as it used to be that proved reserves will ultimately be discovered on the property. Twenty-five percent of the original cost, $2,500, is charged to a loss account in the period that loss of confidence occurs. At the same time $2,500 is credited to an allowance account which is then subtracted from the carrying balance of the Unproved Properties account for balance sheet purposes.

It is important to determine whether or not the impaired property is a large or a small cost center when compared to the remaining unproved properties in the company. If the property is individually significant the impairment is taken on a property-by-property basis. That is, each property is individually assessed and if impairment is evident an individual allowance is set up for that property.

On the other hand, if the property costs are not individually significant (they are not material) the properties may be divided into groups. Impairment may then be taken on a group basis taking into consideration: "The experience of the enterprise in similar situations and information about such factors as the primary lease terms of those properties, the average holding period of unproved properties, and the relative proportion of such properties on which proved reserves have been found in the past."[3]

When a property is abandoned, the firm must compare the balance in the impairment account to the carrying value of the property. If the properties have been impaired on a property-by-property basis, additional loss will need to be taken in abandonment unless the property had been previously impaired 100 percent. On the other hand, if impairment has been done on a group basis, no additional loss should be recognized at the time of abandonment as long as the group impairment allowance is adequate to cover this one abandoned property.

When proved reserves are discovered on an unproved property, the carrying balance of that account is transferred to an account called *Proved Properties*. The significance here is that Proved Properties are amortized over production, whereas Unproved Properties are either not reduced at all or only reduced for impairment. When discovery does take place on a group property, the original historical cost of that property is transferred to Proved Properties. If, however, there was a property-by-property impairment, only the net amount (that is original cost minus any impairment allowance) is transferred to Proved Properties.

EXAMPLE

Assume that the Start Up Oil Company began operations in 1980. During that year the company paid various land owners for the right to go on the properties and conduct seismic studies. Based on those results, Start Up signed several leases; a series of 46 small leases in group A, with lease bonuses of $5,000 per lease, and the larger Buster, Carter, and Delta leases. Notice in Figure 5-1 these lease bonuses are significantly higher. During 1981 G

[3]Statement of Financial Accounting Standards #19, paragraph 28.

and G studies were also conducted and, based on these seismic studies, it was determined that the Buster lease was very unlikely to ever produce any reserves. Therefore, that property has been 100 percent impaired.

EXAMPLE

		GROUP A*	BUSTER LEASE	CARTER LEASE	DELTA LEASE
1980	Shooting Rights: $150,000				
	G and G Costs: $180,000				
1981	Lease Bonuses	$230,000	$175,000	$265,000	$ 80,000
	G and G Costs	20,000	10,000 (100% Impairment)	10,000	10,000
1982	Drilling Results	5 Leases Proved Reserves	None	None	Dry Holes (75% Impairment)
	Delay Rental			$ 6,000	
1983	Drilling Results	6 Leases Proved Reserves 5 Leases Abandonded	None	Dry Holes (No Impairment)	Proved Reserves

For Group A, assume 10% of total leases, both proved and unproved, are impaired each year.

Figure 5—1

Drilling began in 1982: five of the leases on group A were determined to have proved reserves. There was no drilling on the Buster lease because it was impaired, and at this time no drilling was authorized on the Carter lease. Only dry holes were drilled on the Delta lease and that property in 1982 is deemed to be 75 percent impaired. Past history reveals that for leases similar to those in group A, ten percent of the total leases (that means all leases, both proved and unproved) are impaired each year. It is necessary to pay a delay rental on the Carter lease to keep it active. Since the firm has no confidence that the Buster lease will ever produce proved reserves, a delay rental is not paid during 1982, and consequently the provisions of the lease called for the working interest to terminate; therefore, it is now considered abandoned.

In 1983 six additional leases on group A are proved, five of those leases are abandoned, and an additional dry hole is drilled on the Carter lease with results not adequate to require impairment. An important point here is that a single dry hole on a lease, or in some cases even a series of dry holes, does not necessarily mandate impairment. In spite of the previously recognized 75 percent impairment on the Delta lease, in 1983 the firm discovered proved reserves. Impairment does not necessarily mean that proved reserves will not be discovered on a lease.

SUCCESSFUL EFFORTS TREATMENT

Refer to Figure 5-2. Notice that all of the shooting rights and G and G costs in 1980 are charged to an expense. In 1981 all of the lease bonuses are placed in the account called Unproved Properties. The 1981 G and G costs are also charged to an expense. Ten percent of the original cost of the leases in group A is charged to a loss and the appropriate impairment allowance is taken. The carrying balance for lease A would now be equal to $207,000. All of the acquisition costs of lease B are written off to a loss because this property is now considered to be 100 percent impaired. Since legal title to the Buster property will carry over into the next year the impairment allowance is used in this case rather than writing the loss off directly to Unproved Properties.

In 1982, an additional ten percent impairment is taken on group A. Notice that the engineers had determined that ten percent of total leases are impaired each year; this is not simply ten percent of the remaining leases. In addition, the acquisition costs of the five leases which now have proved reserves is transferred from Unproved Properties to Proved Properties. By this time the lease rights on the Buster lease have expired, so the acquisition cost is written off. Since this abandoned property has already shown a 100 percent impairment on the financial statements, we merely need to remove the amounts already on the books. We do this by reducing the impairment allowance in the Unproved Properties account by the original cost of the lease. The delay rental for the Carter lease must be shown as an expense and 75 percent of the original cost of the Delta lease is charged off to a loss to establish the individual impairment allowance for that property.

SUCCESSFUL EFFORTS

		Account	Debit	Credit
1980		Exploration Expense	330,000	
		Cash		330,000
1981		Unproved Properties	750,000	
		Cash		750,000
		Exploration Expense	50,000	
		Cash		50,000
	(A)	Loss—Impairment	23,000	
		Impairment Allowance (Group A)		23,000
	(B)	Loss—Impairment	175,000	
		Impairment Allowance (B)		175,000
1982	(A)	Loss—Impairment	23,000	
		Impairment Allowance (Group A)		23,000
	(A)	Proved Properties	25,000	
		Unproved Properties		25,000
	(B)	Impairment Allowance (B)	175,000	
		Unproved Properties		175,000
	(C)	Delay Rental Expense	6,000	
		Cash		6,000
	(D)	Loss—Impairment	60,000	
		Impairment Allowance (D)		60,000
1983	(A)	Loss—Impairment	23,000	
		Impairment Allowance (Group A)		23,000
	(A)	Proved Properties	30,000	
		Unproved Properties		30,000
	(A)	Impairment Allowance (Group A)	25,000	
		Unproved Properties		25,000
	(D)	Proved Properties	20,000	
		Impairment Allowance (D)	60,000	
		Unproved Properties		80,000

Figure 5—2

In 1983 an additional ten percent impairment allowance is taken on group A and the original cost associated with the six new proved properties is transferred to the proved properties account. Since we have abandoned five of those properties the cost of those properties is removed from the Unproved Properties account as well as the impairment allowance account. Clearly there is an adequate balance in that account for these abandonments. Nothing further need be done with the Buster lease. The Carter lease, even though it has shown dry holes, has no adjustment to the acquisition cost. However, the Delta property now has proved reserves. Unfortunately, we have previously taken impairment on that lease and so we may transfer to the Proved Properties account only the unimpaired balance of $20,000. The original $80,000 is removed from the Unproved Properties account and the $60,000 impairment is removed from that account.

FULL COSTING TREATMENT

The accounting for acquisition cost and full costing is much, much simpler as illustrated in Figure 5-3. All of the costs are simply capitalized as they are incurred. This capitalization is, of course, accomplished only within the confines of the cost center capitalization limit as discussed in Chapter 8. In 1980, the costs of shooting rights and G and G studies are capitalized. In 1981, all the lease bonuses and all the G and G costs are capitalized. In 1982, the costs of the delay rental is capitalized. It should be noted here that other costs were incurred in exploring these leases. These costs will be examined in more detail in Chapter Six.

FULL COSTING

1980	Capitalized Cost Center	330,000	
	Cash		330,000
1981	Capitalized Cost Center	800,000	
	Cash		800,000
1982	Capitalized Cost Center	6,000	
	Cash		6,000
1983	*[No Entry for Acquisition Costs]*		

Figure 5–3

Chapter 6

Accounting for Exploration and Development Costs

As a general definition we can say that *exploration costs* are all those costs incurred in order to decide whether or not proved reserves exist on a given property. Once that doubt has been resolved and it has been determined that proved reserves do exist, everything spent in order to place the property into producing condition is considered to be a *development cost*.

FORMS OF EXPLORATION COSTS

We have already dealt with two types of exploration costs: the shooting rights and the seismic studies performed prior to acquisition of the property. Remember these costs do not, by themselves, find proved reserves, only formations that may have reserves. Exploration costs will also include all seismic studies, geological and geophysical costs incurred after acquisition.

Another form of exploration cost is the delay rental. The delay rental expense is incurred solely to give the company one more year to drill an exploratory well. Since it does not grant additional property rights to the lessee, it is properly considered to be an exploration cost.

Test well contributions paid to another party to drill a well close to our property are also considered exploration costs. These contributions usually are paid to a lessee who holds a lease contiguous or adjacent to a lease owned by another company. Neither party wishes to incur the entire cost of drilling a well to test a formation which geological and geophysical studies show to be on both properties; therefore, it is a frequent practice for one party to drill a well on his property and the other party to contribute to the cost of that well. No mineral rights exchange hands, as only information is involved in this case. There are two common forms of test well contribution. The bottom hole test well contribution is a situation where the party drilling a well receives a contribution from the other party if the well is drilled to an agreed upon depth. Payment occurs whether or not reserves are discovered.

In a dry hole test well contribution the payment is only made if the well is drilled to the agreed upon depth and the hole is plugged and abandoned as a dry hole. In this case the only risk shared is that of a dry hole.

The most expensive exploration cost in most cases is the exploratory well. An *exploratory well* is defined as a well which is not a *development well*, a *service well*, or a *stratigraphic test well*. Let us look for a moment at each of these other three types of wells.

A development well is essentially a well drilled in a stratigraphic area known to be productive and to a stratigraphic depth known to be productive. That is, it is drilled into a formation where proved reserves have already been demonstrated to exist. The purpose of drilling a development well is to produce known reserves.

A service well is generally drilled to support production operations. It may be drilled as a source of water for injection into water flooding projects, or it may be drilled to dispose of BS and W which is produced on the property. In short, all wells drilled to support the production operations for primary, secondary, or tertiary production are considered to be service wells.

Stratigraphic test wells are drilled in order to gather stratigraphic or geological information only. They are usually small bore wells drilled primarily to obtain core samples. It is not their purpose ordinarily to find and produce reserves, but only to gain information.

Exploratory wells, then, are the only wells that remain. Generally speaking, exploratory wells are drilled with the intent of finding proved reserves, but only in those areas where reserves are not already known to exist.

All the costs associated with an exploratory well are considered to be an exploratory expense up to the time it is determined whether or not the well has found proved reserves. If it is determined that no proved reserves are discovered, then the well is plugged and abandoned. These plugging and abandoning cost are considered to be an exploratory cost.

However, if the well is determined to have proved reserves then all costs incurred to complete the well are considered to be development costs.

FORMS OF DEVELOPMENT COSTS

Development costs are all those costs incurred after the property has been determined to have proved reserves up until the time the property is capable of producing. There are several forms of these costs.

The costs necessary to complete the discovery well — that is to case it, cement it, perforate it, fracture it, and install production equipment in the well after proved reserves are determined — are all development costs.

Since most formations cannot be produced from a single discovery well it is necessary to place additional wells into the formation for maximum recovery. All of these wells are considered to be development costs, whether the wells are productive or dry holes. These wells are commonly referred to as *step out wells* because the wells step out from the original discovery well until they no longer find adequate reserves to justify their completion. Eventually, the stepping out process will go far enough out from the center point of the formation, that they are drilled into a formation which does not produce adequate reserves to justify completing these wells. These wells are then plugged and abandoned as dry holes. Approximately one out of four discovery wells is ultimately determined to be a dry hole.

The remaining surface equipment which is installed is also considered to be a development cost. This equipment includes the christmas tree or pumping unit, gathering lines, treatment facilities, tanks, lact meters and any other equipment installed on the lease. In short, the surface equipment includes everything to get the product up out of the ground, treat it, measure it, store it, and make it ready for market. Once the product leaves the lease the costs are no longer considered to be development costs, but become transportation costs.

GENERAL ACCOUNTING RULES FOR EXPLORATORY COSTS

For successful efforts companies all exploratory costs except those which result in specific new proved reserves should be expensed. Included will be shooting rights, G and G costs, delay rentals, and test well contributions.

For successful efforts companies all exploratory wells are first capitalized. When the drilling process has been completed the well is evaluated. If no reserves are discovered then the well is plugged and abandoned and the costs of the well are reclassified as a dry hole expense. If reserves are discovered but it cannot yet be determined if there are adequate reserves to qualify as proved reserves, the costs of drilling the well may be held as an asset for one year after the end of drilling. If, at the end of that year, the reserves still cannot be reclassified as proved reserves the costs of the well are charged to dry hole expense. If this happens and the well is subsequently determined to have proved reserves the costs carried forward are only those costs incurred after the one year time limit caused the costs of the well to be written off. If a major capital expenditure is required before the development wells are drilled this one year limitation can be bypassed.

For full costing companies, all costs associated with property exploration are to be capitalized within the capitalized cost center. This will include all of the shooting rights, G and G costs, delay rentals, and all drilling costs whether the wells are productive or non-productive. Of course this capitalization process is limited by the capitalization limit as described in Chapter Eight.

GENERAL ACCOUNTING RULES FOR DEVELOPMENT COSTS

For both successful efforts and full costing companies, all costs incurred to develop a property, that is to bring it to a state of production readiness after it is determined to have proved reserves, are capitalized within the appropriate cost center. Costs of all development wells, both successful and dry holes, should be capitalized. Remember, of course, for a successful efforts company this cost center will be relatively small, while for a full costing company the cost center ordinarily will be the United States.

EXAMPLE

This example, as shown in Figure 6-1, will be a continuation of the pre-acquisition and acquisition cost example in Chapter Five. We will alter some of the facts to illustrate the accounting for exploration and development costs.

Assume that three leases, Buster, Carter, and Delta, are acquired in 1981, after preliminary seismic studies in 1980. Let's also assume that no drilling ever takes place on the Buster lease, but that one exploratory well is drilled on the Carter lease and two are drilled on the Delta lease in 1982. The Carter well is a dry hole. Reserves are discovered on the Delta lease, but they cannot yet qualify as proved reserves.

EXAMPLE

		BUSTER LEASE	CARTER LEASE	DELTA LEASE
1980	Shooting Rights: $150,000			
	G and G Costs: $180,000			
1981	G and G Costs	$10,000	$ 10,000	$ 10,000
1982	Drilling Costs	None	$300,000	$600,000
	Results		Dry Hole	Indeterminate
1983	Drilling Costs	None	$400,000	None
	Results		Proved Reserves	Indeterminate
	Development Costs		$800,000	
1984	Results			Proved Reserves
	Development Costs			$1,800,000

Figure 6—1

Assume that in 1983 an additional well is drilled on the Carter lease and proved reserves are discovered. That well is completed and one additional development well is drilled. Assume that the reserves on the Delta lease are still indeterminate and that nothing further is spent on that property in 1983. In 1984 assume the price of crude oil has increased to a point where the reserves on the Delta lease are now feasible. At that point they are determined to be proved reserves.

SUCCESSFUL EFFORTS TREATMENT

In 1980 the shooting rights and preliminary seismic studies are charged to expense. In 1981 the detailed seismic studies on the leases are also charged to expense. See Figure 6—2.

All three wells drilled in 1982 are initially capitalized in the account titled *Uncompleted Wells and Equipment*.[1] The significance

[1]**This account is sometimes referred to as *Uncompleted Wells and Related Equipment and Facilities*. The shorter term is used here for convenience sake. Also used is the term *Wells in Progress*.**

of this account is that it is a capitalized asset, but it is not amortized over production.

The one well on the Carter lease is determined to be a dry hole and is plugged and abandoned, and its costs are charged to a dry hole expense. At year end 1982, the costs that relate to the Delta lease wells are still held in the account Uncompleted Wells and Equipment, pending the determination of whether or not proved reserves have been found.

In 1983, an additional well is drilled on the Carter lease, and these costs are capitalized in the account titled Uncompleted Wells and Equipment. When the well is determined to have proved reserves the costs are transferred to *Wells and Equipment* along with the costs needed to complete and develop the property. Therefore, the entire cost will be amortized over production. By year end 1982, the Delta lease is still undetermined; that is, we still do not know if we have proved reserves.[2] Since one year has passed since the end of drilling on the Delta lease, the cost of the two exploratory wells must be written off to a dry hole expense.

Assume that the costs spent on the Delta lease in 1984 consist of completing the two exploratory wells and installing additional development wells and related surface equipment. These are the only exploratory and development costs capitalized on this lease. We do not go back and reclassify the $600,000 written off as dry hole expense, even though it ultimately found proved reserves.

FULL COSTING TREATMENT

Each year the total amount spent on exploratory efforts on all leases is capitalized in the single capitalized cost center. Notice that we make no determination for accounting purposes whether wells are exploratory or development, whether or not they find proved reserves, and whether or not properties are abandoned.

As usual, the costs can be capitalized to the extent that they do not exceed the capitalization limit as discussed in Chapter Eight. See Figure 6—3.

[2]Assume that a delay rental was paid to keep this lease in effect.

SUCCESSFUL EFFORTS

Year		Account	Debit	Credit
1980		Exploration Expense	330,000	
		Cash		330,000
1981		Exploration Expense	30,000	
		Cash		30,000
1982		Uncompleted Wells and Equipment	900,000	
		Cash		900,000
	(C)	Dry Hole Expense	300,000	
		Uncompleted Wells & Equipment		300,000
1983		Uncompleted Wells and Equipment	400,000	
		Cash		400,000
	(C)	Wells and Equipment	1,200,000	
		Uncompleted Wells & Equipment		400,000
		Cash		800,000
	(D)	Dry Hole Expense	600,000	
		Uncompleted Wells & Equipment		600,000
1984	(D)	Wells and Equipment	1,800,000	
		Cash		1,800,000

Figure 6–2

FULL COSTING

		Debit	Credit
1980	Capitalized Cost Center	330,000	
	Cash		330,000
1981	Capitalized Cost Center	30,000	
	Cash		30,000
1982	Capitalized Cost Center	900,000	
	Cash		900,000
1983	Capitalized Cost Center	1,200,000	
	Cash		1,200,000
1984	Capitalized Cost Center	1,800,000	
	Cash		1,800,000

Figure 6–3

Chapter 7

Accounting for Production and Revenue

In this chapter we shall complete the process and discuss the accounting techniques for producing the product, treating it, and recording the revenue as the product is sold.

PRODUCTION COSTS

Recall that all costs incurred to lift the product to the surface, treat it, and make it ready for sale, are referred to as production costs. These costs, also called lifting costs, include all the operating costs of the lease, plus any normal routine repairs and maintenance. In addition, there are three categories of taxes which are included as production costs.

The first are property taxes, charged by local authorities for the value of the facilities installed. Second are *severance taxes*, charged by the state in which the product is produced. The state charges these taxes because the oil company has severed a natural resource from the state. These taxes sometimes are also referred to as *production taxes*. All parties who have an economic interest in the reserves, including both the working interest and the royalty interest must pay the severance tax.

The final category of taxes is the crude oil windfall profit tax. This tax, enacted by the 1980 Crude Oil Windfall Profit Tax Act, must be paid by most producers of crude oil and condensate. Generally speaking, the tax is higher when applied to oil that is produced from older wells, and lower when applied to oil that is produced from certain newer wells and other special category wells, such as stripper wells and heavy oil.

In most cases the severance tax for all parties involved is collected by the producer of the oil, who then remits the total amount to the state and sends the balance of the revenue, minus the severance tax, to the royalty owner and other joint interest owners. In many cases the crude oil windfall profit tax is withheld by the first purchaser of the oil, the pipeline or the refinery that purchases the oil from the producer. Only the balance is remitted then to the producer who sends the net amount on to the royalty owner and joint interest owners.

BASIC ACCOUNTING TREATMENT

For both full costing and successful efforts companies, production costs are usually expensed in the year incurred.[1] The producing company should include in its production cost expense for the period only those severance taxes and windfall profit taxes which relate to the working interest owned by that producing company. Taxes which have been collected for other entities such as the severance tax and the royalty interest, must be shown as a liability until paid to the state.

WORKOVERS

A *workover* is a situation where an existing producing well is temporarily shut down and some form of maintenance activity is applied to that well. This could include repair of damaged equipment, acidizing of perforations which have been clogged by parafin, a refracturing of the well, or any one of numerous other reasons for entering the bore. As a general rule, only the workover costs which are associated with existing completion intervals, that is, existing formations that are already producing, should be charged to expense. Workover costs incurred to bring other formations into production from an existing well should be treated as in the previous chapters. For example, refer to Figure 7-1. Assume that the well was drilled first through formation A, which proved to be marginally productive but was not completed. The well was extended to formation B, which was completed and is now producing. Any activity that maintains or increases production from formation B is considered a workover cost to be immediately expensed. Suppose that prices have risen and the company plugs back to formation A, completes it, perforates it, and brings it into production from the same bore. This would be capitalized for both full costing and successful efforts companies.

[1]Under certain circumstances the lifting cost may be inventoried as a product cost for successful efforts companies at year end. This is a requirement of Statement of Financial Accounting Standards Number 19, which requires that the production costs shall become part of the cost of the product produced. On the other hand, full costing companies must charge the production cost to expense as incurred.

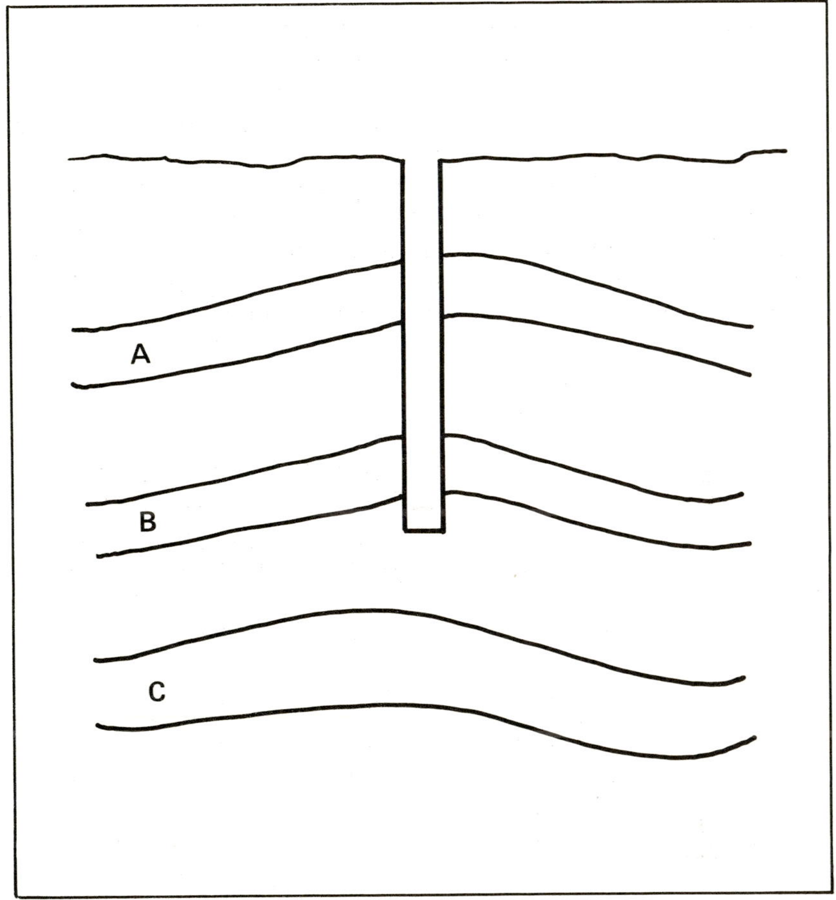

Figure 7–1

On the other hand, assume the company decided to extend the bore to formation C to test it for potential proved reserves. In a full costing company, the cost of this extension will be capitalized in any event. In a successful efforts company this extension will be considered an exploratory well and will be capitalized if successful, but expensed if a dry hole.

EXAMPLE

We can illustrate the accounting for these production costs and the distribution of revenue with the following circumstances.

Assume the Black Gold Oil Company owns the working interest in the Sure Thing producing property. Land Owner has a royalty interest of one-eighth of total production, Also Ran has an overriding royalty interest of one-eighth of total production. During May 1981, production was 3,000 barrels of crude oil; the average sales price was $32 per barrel, average windfall profits tax was $8 per barrel, and state severance tax was 10 percent of gross sales price. Lifting costs were $4 per barrel. Also assume that Black Gold incurred workover costs of $2,000 to improve production from the existing completion interval. Figure 7-2 shows the distribution of revenue from this set of circumstances.

		WORKING INTEREST	ROYALTY INTERESTS	
	TOTAL	BLACK GOLD	LAND OWNER	ALSO RAN
Barrels	3,000	2,250	375	375
Revenue (Gross)	96,000	72,000	12,000	12,000
Lifting Costs	(12,000)	(12,000)		
Windfall Profit Tax	(24,000)	(18,000)	(3,000)	(3,000)
Severance Tax	(9,600)	(7,200)	(1,200)	(1,200)
Workover Costs	(2,000)	(2,000)		
Net Proceeds From Revenue	48,400	32,800	7,800	7,800
	100%	68%	16%	16%

Figure 7—2

Notice that of the 3,000 barrels, six-eighths are those of the working interest. The two royalty owners each have an interest in one-eighth of the total barrels produced. Gross revenues are calculated for each entity by simply multiplying the total barrels produced by $32 per barrel. Lifting costs are a total cost of the working interest. Under most lease contracts, the royalty owners do not participate in any of the lifting costs except severance and windfall profit taxes.

Each entity must bear its own share of the crude oil windfall profit tax. The total tax is distributed in the same ratio as the barrels of crude oil. The same is true of the severance tax. The workover costs are treated the same as lifting costs; that is, they are born entirely by the working interest owner. Notice that the net proceeds from the month's production are somewhat different in percentage than the ownership of reserves. That is, Land Owner and Also Ran owned 12½ percent of the reserves but each received 16 percent of the net proceeds. Black Gold owns 75 percent of the mineral interest but received only 68 percent of the net proceeds. This is true because the working interest owner must bear all lifting, production, and workover costs. These numbers are adjusted even further when one realizes Black Gold also had to bear the entire cost of acquisition, exploration, and development of the well. When these costs are amortized over the property the percent of total proceeds from the well becomes even less for the working interest owner.

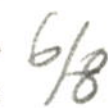

The journal entries as illustrated in Figure 7-3 show the disposition of the distribution of revenue. Notice that the production expense is charged in total to Black Gold Company.

The second entry requires substantial explanation. We will assume in this case that the first purchaser of the crude oil, the pipeline company which purchases the 3,000 barrels from Black Gold, withheld the crude oil windfall profits tax and will remit that tax to the U.S. Treasury. That means that Black Gold received cash equal to the revenue minus the windfall profit tax. Black Gold also had to bear windfall profit tax expense and severance tax expense as indicated in the entry.

Even though each royalty owner had gross revenues of $12,000, they both net $7,800 after the two taxes are subtracted. Therefore, the total amount due the two royalty owners is equal to the $15,600 indicated in the entry. Black Gold will need to pay severance taxes to the state for all interests involved, so the liability Severance Taxes Payable is for all severance taxes on total production. The remaining $72,000 revenue is the gross revenue for the working interest owner.[2]

[2]It is only coincidence in this example that the cash received ($72,000) is equal to the revenue for the working interest.

WORKING INTEREST			ROYALTY INTERESTS		
BLACK GOLD			**LAND OWNER AND ALSO RAN**		
Production Expense	12,000		No Entry		
Cash		12,000			
Cash	72,000				
Windfall Profits Tax Expense	18,000		No Entry		
Severance Tax Expense	7,200				
Royalties Payable		15,600			
Severance Tax Payable		9,600			
Revenue		72,000			
Royalties Payable	15,600		Cash	7,800	
Cash (Land Owner)		7,800	Windfall Profits Tax Expense	3,000	
Cash (Also Ran)		7,800	Severance Tax Expense	1,200	
Workover Expense	2,000		Revenue		12,000
Cash		2,000			

Figure 7—3

When the royalty checks are sent, Land Owner and Also Ran will record the cash received and their own tax expenses for that period. This usually takes place approximately 30 days after the end of production. In most cases, at year end the royalty owners will need to accrue or estimate the amount of revenue due them that they have not yet received.

Finally, notice that the workover expense is shown as a total expense of Black Gold production.

Chapter 8

Accounting for Amortization

The previous three chapters have demonstrated how we account for current expenditures, and whether those expenditures are capitalized or expensed. At the end of each period, the oil and gas producing company must also charge, on a units of production basis, revenue with a proportionate share of previously capitalized cost. This process is called *amortization*. In this chapter we shall examine the amortization process and its accounting. We will first examine the basic amortization process, then look at the specific adjustments needed under both full costing and successful efforts, and then follow with a comprehensive example.

THE BASIC AMORTIZATION PROCESS

Since amortization for oil and gas producing companies is done on a per barrel or units of production basis, all amortization efforts are devoted toward developing a rate or a dollar amount per barrel. This is expressed as the fraction illustrated in Figure 8-1. Notice that the end result is a dollar amount per barrel for each cost center. That means a full costing company will have one amortization rate for the entire United States, plus another rate for

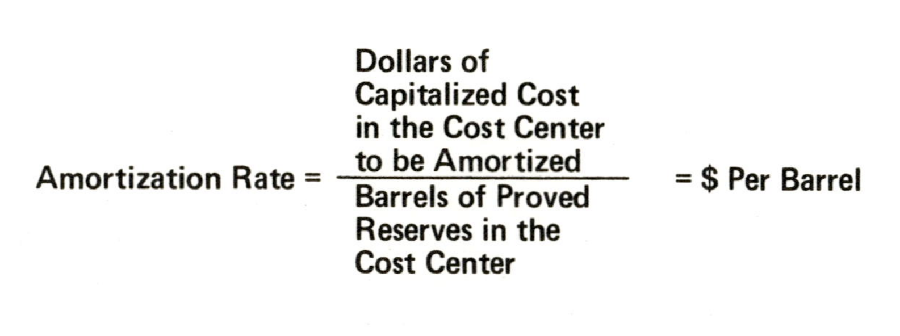

$$\text{Amortization Rate} = \frac{\text{Dollars of Capitalized Cost in the Cost Center to be Amortized}}{\text{Barrels of Proved Reserves in the Cost Center}} = \text{\$ Per Barrel}$$

Figure 8—1

each foreign geographic cost center. A successful efforts company will compute amortization rates for each of its many cost centers. As we have noticed before this is an averaging process for the full costing companies.

INPUTS FOR THE RATE

A successful efforts amortization rate includes three items as illustrated in Figure 8-2. These items include those acquisition costs which are related to properties that have been deemed to have proved reserves. Acquisition costs of properties which have been impaired, and acquisition costs of unproved properties which are still in doubt, are not included in the computation of this rate, but the costs of exploratory discovery wells are included. Other exploration costs such as geological and geophysical costs, delay rentals, test well contributions, and exploratory dry holes, are not included in the computation of this rate. All development costs, including development dry holes, are included.

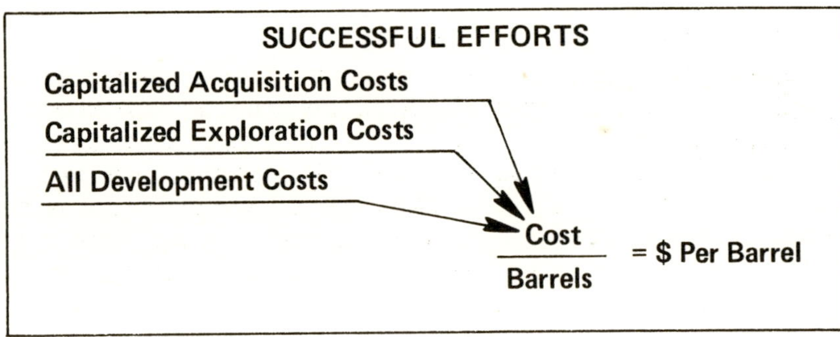

Figure 8—2

The full costing amortization fraction is much broader in scope as indicated in Figure 8-3. In this case all the costs to acquire properties, whether proved or unproved, successful or unsuccessful, are included. In the same light, all costs to explore those properties, whether successful or unsuccessful, are included, as are all development costs. It should be obvious that the full costing rate will generally be much higher than the average successful efforts amortization rate.

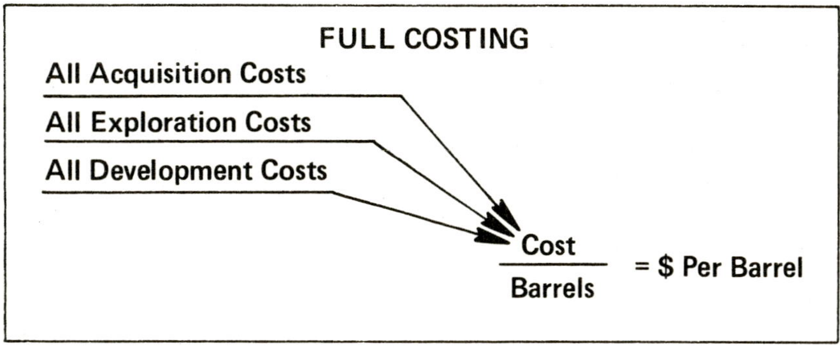

Figure 8—3

ADJUSTMENTS

This fraction is a simplification of the actual process. There are several adjustments which must take place to both the numerator and denominator in determining each specific amortization rate.

SALES, RETIREMENTS, ETC.

When a property has been sold, abandoned, conveyed or retired, the cost associated with that property must be removed from the numerator of the cost center amortization fraction. This is true for both full costing and successful efforts. Frequently when a property is sold by a successful efforts company it represents an entire cost center, and, consequently, that amortization rate is eliminated. Sometimes, however, only a portion of the cost center is sold or conveyed. In this case a reduction of both the numerator (the dollars involved) and the denominator (the volume of product), must be eliminated from the fraction. For full costing companies, both the costs and the barrels involved must be removed from the overall cost center when such a sale or conveyance takes place.

PLANT REMOVAL AND SITE RESTORATION COSTS

Both successful efforts companies and full costing companies are required to include a consideration of costs to dismantle, abandon, and restore a production location in the amortization rate. This would include a cost of dismantling any production facilities, including pipelines, offshore platforms, and treatment facilities, as well as the cost to restore the land to its original condition. These costs frequently are nominal on onshore locations; however, more and more production is beginning to take place in hostile environments, such as the north slope of Alaska and deep water offshore locations. In these situations the environment is disturbed by the exploration and development process, and it is necessary for the company to incur very substantial costs in order to restore the environment after production ceases. Since the company must incur this obligation when production begins, the obligation is a part of the cost of the production facility; therefore, we must consider this cost in computing the amortization rate. The matter of inflation becomes very important here if, for example, an offshore platform is to be removed in 20 years; the actual dollars to be spent are going to be much greater than if we were to remove the platform in today's economy. It is generally held that the plant removal and site restoration provision for amortization should be included at today's cost, rather than at what the inflated dollars will be when the restoration actually takes place. The example will illustrate that no general entry is made at present to place the plant removal and site restoration costs on the balance sheet. This is only a work sheet computation to increase the rate per barrel. Therefore, the removal and restoration costs enter the balance sheet only indirectly through accumulated amortization. In all probability, the FASB or the SEC will soon require additional accounting disclosures for these obligations.

FUTURE DEVELOPMENT COSTS

Full costing companies are required to include in their numerator an estimation of what they will spend in the future to develop proved reserves which are currently undeveloped.[1] For all reserves which qualify as proved undeveloped, the company must estimate in present dollars the costs which are likely to be incurred to develop those properties to the point where they are capable of producing.

AMORTIZATION DELAY ITEMS

There are also two circumstances when costs which have been capitalized in a full costing company's capitalization cost

[1]Refer to Chapter Two for a definition of developed versus undeveloped reserves.

center are omitted from the amortization base. In these cases the costs are not removed from the balance sheet; they are simply temporarily eliminated from the computation of the amortization rate.

In the first case, the company has incurred very substantial costs in a particular project. These costs are material to the total cost of the entire cost center. In addition, the company must have not yet determined that proved reserves exist in the project, but the company must have determined that impairment has not yet taken place. In other words, there is still substantial doubt about the ultimate success or failure of this major project. Since there is still doubt, the reserves cannot be included in the denominator because they have not yet been classified as proved reserves. If the dollars of cost were included in the numerator without putting the reserves in the denominator, the rate would be too high. Thus, the full costing company can withhold their costs from rate determination until doubt is resolved.

In the second case, the full costing company has determined that proved reserves exist, but is unable to determine the extent of those reserves until additional wells have been drilled. In this situation, a producing formation has been confirmed, but the exact size of that formation will depend on additional wells. We could include in the denominator only the proved reserves confirmed by existing wells. The SEC has provided that we can delay putting costs associated with this project into the amortization rate until such time as we have completed drilling the wells to determine total reservoir size.

REVISION OF RESERVES ESTIMATES

Perhaps the only thing we know for sure about the beginning estimates of the recoverable proved reserves present in a new formation is that the estimate is wrong; that is, it will be revised. This is particularly true in the early years of the life of the reservoir. Both the FASB and SEC have provided that the oil and gas producing company will, on a periodic basis and at least annually, revise the estimates of proved reserves. The easiest way to accomplish this is to add to the ending estimate of proved reserves production which has taken place during the period. This will give the best available estimate of beginning proved reserves. In most cases, this revised beginning proved reserve estimate will be different from last period's ending reserve estimate. The accounting pronouncements provide that this estimate will not be applied to previous years' income; that is, we will not go back and adjust our amortization expenses or accumulated amortization in past years. Instead, the adjustment is called a prospective adjustment. We simply take the unamortized cost and divide it by the remaining barrels. This spreads any error in our rate over the future unproduced barrels. Since most companies issue quarterly statements, a revision of reserve estimates on a quarterly basis is appropriate if there is sufficient engineering evidence to indicate that the reserve estimates previously used have been incorrect.

SUCCESSFUL EFFORTS DENOMINATORS

There is no provision in the pronouncements for successful efforts companies to include future development costs in the amortization rate. This causes a conceptual problem because the denominator of the fraction includes reserves which are proved but undeveloped; however, the numerator does not allow for the costs associated with developing those reserves. Therefore, we must make the following adjustment for successful efforts companies. Acquisition costs which have been capitalized are amortized on a basis of all proved reserves, both developed and undeveloped. Capitalized exploration costs and capitalized development costs, however, are amortized on a basis of proved developed reserves only; proved undeveloped reserves are not included in that denominator.

Cost centers which do not have any proved undeveloped reserves will be able to use a single amortization rate, under successful efforts, for that property. This is true because, in this case, the quantity of all proved reserves is equal to proved developed reserves. However, those cost centers which include proved reserves which have not yet been developed will need two amortization rates, one for acquisition cost (based on all proved reserves, both developed and undeveloped) and another for capitalized exploration and development cost (based on proved developed reserves only).

CAPITALIZATION LIMIT

The lower of cost or market rule may be applied when establishing the amortization rate.[2] If the fair market value (or the

[2]There is no specific requirement for a cost center-by-cost center capitalization limit in successful efforts. However, it is not inappropriate to apply the lower of cost or market rule on a cost center-by-cost center basis.

net realizable value) of a cost center is less that the net capitalized cost (actual historical cost less accumulated amortization) the value of that cost center must be written down to the net realizable value.

There is no specific provision for determining the net realizable value for successful efforts companies. This determination is left to the integrity and judgment of the oil and gas producing companies.

However, the SEC has provided a very specific valuation method for full costing companies. These companies must compare an amount equal to the Reserve Recognition Accounting value[3] with the net capitalized value. The net capitalized value should not include future plant removal site restoration costs or future development costs, but should be limited to the actual historical cost of the capitalized cost center minus any accumulated amortization.

The writedown will be a permanent writedown; that is, it should be a reduction of the capitalized cost center, with an offsetting expense to the current year's income. Should the market value go back up in future years the writedown is not reversed. The writedown is on a cost center-by-cost center basis so that there will be a comparison for each individual successful efforts cost center. However, full costing companies which have operations only in the United States will compare only the total market value of all properties to the total historical cost of that cost center.

EXAMPLE

The following comprehensive example illustrates these adjustments.

RESERVE DATA

This example assumes the company owns three leases. Lease A has developed reserves of both oil and gas and, in addition, some undeveloped reserves of oil. Lease B has reserves of oil only and these reserves are all developed. Lease C is acquired during 1980, and explored and abandoned in the same year; no proved reserves were found on lease C. The data in Figure 8-4 shows the beginning and ending reserve estimates as well as the 1980 pro-

[3]See Chapter Ten for a discussion of this technique.

duction. Notice that the estimates of reserves on lease A increased for both developed and undeveloped reserves. However, the reserve estimates on lease B decreased. They will both need to be revised in the amortization computations.

EXAMPLE RESERVE DATA

		LEASE A		LEASE B	LEASE C
		Developed	Undeveloped	All Developed	
January 1, 1980					
Reserve Estimate	Oil	80,000 bbl	200,000 bbl	150,000 bbl	–0–
	Gas	300,000 mcf	–0–	–0–	–0–
1980 Production	Oil	8,000 bbl	–0–	16,000 bbl	–0–
	Gas	24,000 mcf	–0–	–0–	–0–
December 31, 1980					
Reserve Estimate	Oil	81,000 bbl	220,000 bbl	124,000 bbl	–0–
	Gas	294,000 mcf	–0–	–0–	–0–

Figure 8—4

FINANCIAL DATA

The financial data for the example is given in Figure 8-5. The book values refer to net book values; that is, actual historical costs minus any accumulated amortization. For successful efforts the data is divided into acquisition cost and exploration and development cost. This is necessary because lease A has proved undeveloped reserves. Consequently, two amortization rates must be developed for that lease. The full costing book value is significantly higher than the successful efforts book value because more costs have been included. The plant removal and site restoration costs are assumed to be a present cost, as are the future development costs. The $200,000 spent on lease C is equal to the amount to acquire, explore, and abandon that property. These costs are all expensed under successful efforts, since no proved reserves were discovered in that cost center. They are, however, capitalized under full costing. The Reserve Recognition Accounting value is given for purposes of determining the capitalization limit under full costing.

EXAMPLE FINANCIAL DATA

	LEASE A	LEASE B	LEASE C
January 1, 1980			
Net Book Value			
SE Acquisition	$ 71,000	$ 60,000	–0–
Exploration and Development	1,388,000	280,000	–0–
FC	2,846,000	532,000	–0–
Estimated Plant Removal and Site Restoration	60,000	80,000	–0–
Estimated Future Development Costs	800,000	–0–	–0–
Current Expenditures	–0–	–0–	$200,000
RRA Value	$3,900,000	$700,000	–0–

Figure 8–5

SUCCESSFUL EFFORTS AMORTIZATION

Figure 8-6 shows the computation of the numerator and denominator for the successful efforts amortization rate. Notice that the costs related to the acquisition of the property are computed separately from the costs for the exploration and development of property in lease A. This is necessary because lease A contains undeveloped proved reserves, and we must calculate two separate amortization rates for that lease. Lease B does not have proved undeveloped reserves, so we can combine all costs into one rate. In lease A the plant removal and site restoration costs are included with the exploration and development costs, because the environment typically is not disturbed when the property is acquired.

The denominator is computed by adding the best estimate of ending proved reserves to the volumes produced during the period. This gives the revised estimate of beginning reserves. The gas is converted to oil at the rate of six mcf of gas equals one barrel of oil.[4] We convert gas to oil here because, in all cases in this example, oil is the dominant product. For lease A we consider all proved reserves in computing the amortization rate for acquisition costs. However, we consider only proved developed reserves in computing the rate for capitalized exploration and development costs. We do not have to compute two rates on lease B since all reserves on that lease are proved developed.

SE AMORTIZATION

	Acquisition	Exp. & Dev.	All Costs
Numerator			
Net Book Value	$ 71,000	$1,388,000	$340,000
Plant Removal & Site Restoration	–0–	60,000	80,000
Total Costs	$ 71,000	$1,448,000	$420,000
Denominator	**All Proved**	**Developed**	**All Proved**
Ending Reserves			
Oil	301,000 bbl	81,000 bbl	124,000 bbl
Gas (equivalent)	49,000 bbl	49,000 bbl	–0–
1980 Production			
Oil	8,000 bbl	8,000 bbl	16,000 bbl
Gas (equivalent)	4,000 bbl	4,000 bbl	–0–
Estimated Beginning Reserves	362,000 bbl	142,000 bbl	140,000 bbl

Figure 8–6

[4]Both the FASB and the SEC require conversion to a common unit of measure on the basis of energy content. The normal rate is 1 bbl of crude oil = 6 mcf of natural gas; however, in specific cases, actual energy content ratios may be different and thus different conversion factors used.

The fractions are then solved to determine the rate per barrel in each case as illustrated in Figure 8-7. Since 8,000 barrels of crude oil were produced and 24,000 mcf (equivalent to 4,000 barrels) there is a total of 12,000 equivalent barrels of production on lease A. It is not unusual to have very dramatic differences in amortization rates of different cost centers for successful efforts companies; this is true because there frequently is no relationship between the total dollars spent in a cost center and the total number of barrels of reserves actually discovered.

The amortization entry includes a debit to amortization expense to reduce 1980 income, and a credit to accumulated amortization to reduce the balance sheet carrying value of all of the cost centers.

FULL COSTING AMORTIZATION

No adjustment need be made for the costs capitalized because the total RRA value exceeds the total net book value. Figure 8-8 shows that both plant removal and site restoration costs and future development costs are added to the numerator.

The denominator is the sum of all company proved reserves, both developed and undeveloped. Notice that, here too, gas is converted to oil on a 6 to 1 basis.

The single fraction, as computed in Figure 8-9, applies to all production for the cost center. The full costing amortization rate will generally be higher than a comparable successful efforts rate.

SE RATES

	LEASE A		LEASE B
	Acquisition	**Exp. & Dev.**	**All**
Rate	$\frac{\$71,000}{362,000 \text{ bbl}} = \$.20/\text{bbl}$	$\frac{\$1,448,000}{142,000 \text{ bbl}} = \$10.20/\text{bbl}$	$\frac{\$420,000}{140,000 \text{ bbl}} = \$3.00/\text{bbl}$

1980 Amortization	Lease A:	12,000 bbl X $10.40 =	$124,800	
	Lease B:	16,000 bbl X $3.00 =	$ 48,000	
			$172,800	
Entry	Amortization Expense		172,800	
	Accumulated Amortization			172,800

Figure 8–7

FC AMORTIZATION

	LEASE A	LEASE B	LEASE C	TOTAL
Numerator				
Net BU	$2,846,000	$532,000	–0–	$3,378,000
Plant Removal	60,000	80,000	–0–	140,000
Future Development	800,000	–0–	–0–	800,000
Current Expenditures	–0–	–0–	200,000	200,000
	$3,706,000	$612,000	$200,000	$4,518,000
Denominator				
Ending Reserves				
Oil	301,000 bbl	124,000 bbl	–0–	425,000 bbl
Gas (equivalent)	49,000 bbl	–0–	–0–	49,000 bbl
1980 Production				
Oil	8,000 bbl	16,000 bbl	–0–	24,000 bbl
Gas (equivalent)	4,000 bbl	–0–	–0–	4,000 bbl
Estimated Beginning Reserves	362,000 bbl	140,000 bbl	–0–	502,000 bbl

Figure 8–8

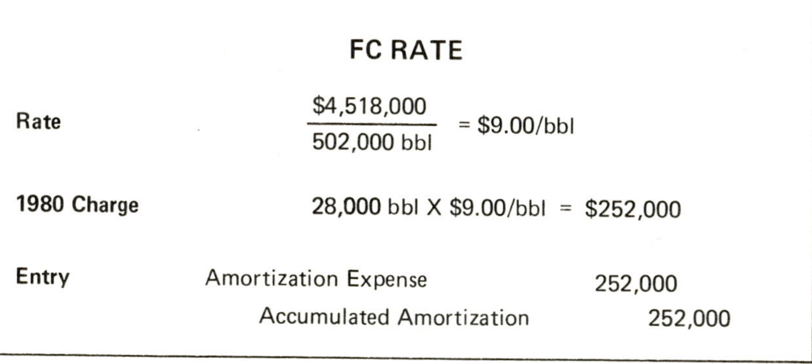

FC RATE

Rate	$\frac{\$4{,}518{,}000}{502{,}000 \text{ bbl}} = \$9.00/\text{bbl}$		
1980 Charge	28,000 bbl X $9.00/bbl = $252,000		
Entry	Amortization Expense	252,000	
	Accumulated Amortization		252,000

Figure 8–9

Chapter 9

Overview of Federal Income Taxation for Oil and Gas Producing Companies

The Federal tax code that relates to oil and gas producing companies has two primary characteristics. It is extremely complex, and it is quite substantially different from the basic rules which are followed for financial accounting for oil and gas producing companies. This chapter will summarize those differences and will compare basic tax accounting rules to the financial accounting rules in an example.

PRIMARY DIFFERENCES

The rules for computing income taxes for oil and gas producing companies vary from financial accounting rules in three basic areas as summarized in Figure 9-1. The differences concern:

1. The decision used to determine whether an expenditure should be capitalized or expensed;
2. The cost center size, and
3. The computation of *depletion* (amortization).

CAPITALIZATION RULES

In successful efforts accounting the primary criteria to determine when costs should be capitalized is whether or not specific proved reserves have been found as a result of that effort. In full costing accounting the primary criteria is whether or not the cost was incurred in pursuit of proved reserves. In tax accounting neither of these rules is applicable.

The basic rule for tax purposes is whether or not a cost is tangible or intangible. For accounting purposes intangible costs generally have been held to be those costs which are related to an asset that does not have a physical existence, such as patents, copyrights, and certain leasehold rights.

For oil and gas tax purposes, a cost is generally held to be intangible for an oil and gas producing company if it does not result

COMPARISON OF ACCOUNTING METHODS

	SE	FC	Tax*
Acquisition Costs	Capitalize if Not Impaired	Capitalize	Capitalize if Not Impaired
G and G Costs	Expense	Capitalize	Capitalize
Delay Rentals	Expense	Capitalize	Expense
Tangible Drilling Costs	Capitalize if Successful Exploration Well or if Develop. Well	Capitalize	Capitalize
Intangible Drilling Costs	Capitalize if Successful Exploratory Well or if Develop. Well	Capitalize	Expense
Surface Equipment	Capitalize	Capitalize	Capitalize
Lifting Costs	Expense	Expense	Expense

**For tax purposes the term "expense" means the item can be deducted from revenue in determining taxable income.*

Figure 9–1

in an asset which has a salvage value. Thus, we have the whole category of *intangible drilling and development costs* (IDC).

Let us look at some specific expenditures in the drilling of a well:

1. The cost of drilling the bore is intangible, because it does not have a salvage value.
2. The cost of cementing the well is intangible.
3. The cost of logging the well is intangible.
4. The cost of perforating and fracturing the well is intangible.
5. Tubing installed to produce the well is tangible, because it could be removed and sold.

6. The cost of the christmas tree is considered to be tangible; however, the cost of installing the christmas tree is held to be intangible.
7. All casing is held to be tangible because, even though it is not likely to be removed, if it were removed it would have a salvage value.
8. All surface equipment including gathering lines, treatment facilities, storage tanks, and LACT meters, among others, are held to be tangible. The cost of installing the surface equipment also is tangible.

We can summarize in this fashion: everything down hole from the christmas tree, which has a salvage value, is tangible; however, the cost of installing those items is held to be intangible. Everything downstream, or on the other side of the christmas tree, is held to be tangible, as are the costs of installing those items.

The important fact here is that all costs which are held to be intangible (classified as IDC) may be deducted in the year in which the costs are incurred regardless of either the results of the well or the life of the well.

This is a major tax benefit since the majority of drilling costs frequently are classified as IDC and can be written off in the year incurred. Because of the time value of money, this has benefit to the company. The total amount of deduction written off over the life of the lease is not greater for tax purposes in this case than it is under financial accounting; however, the deduction occurs very early in the life of the asset, so that taxable income is lower in earlier years and higher in later years. This defers the payment of taxes until the later years of the life of the lease.

COST CENTER SIZE

Recall that for full costing companies the cost center is the entire United States. Most successful efforts companies use a field or reservoir for their cost center. For tax purposes the cost center is considered to be the individual property. This is important because if a property proves to be unsuccessful and can qualify for impairment under the tax rules, all costs associated with that property, even the tangible costs, may be written off in the year impairment is deemed to have occurred. Consequently, the only costs which must remain capitalized and be carried forward as an asset and not immediately deducted against income are those tangible costs associated with leases. Only leases which are either productive or about which substantial doubt still remains, will have these tangible costs.

DEPLETION FOR TAX PURPOSES

In Chapter Eight we discussed amortization for financial accounting purposes. In both full costing and successful efforts companies the original historic costs of those items capitalized were written off against income on a unit of production basis. The controversy between the two methods occurred in what was capitalized, but in both cases the capitalized amount was amortized on a per barrel basis.

Oil and gas companies have two options for amortization of those costs which were capitalized for tax purposes. The first is to follow *cost depletion* as for financial accounting purposes. Those items which were not expensed, essentially those items which do not qualify as IDC, are depleted or amortized on a units of production basis. It should be noted here that in most cases this will be a lower amortization base for tax purposes than for financial statements, because most of the drilling costs will have been deducted in the year incurred as IDC.

The second method is referred to as *statutory* or *percentage depletion*. Congress originally gave an incentive to oil and gas companies to explore for more oil and gas reserves. This incentive took the form of a depletion allowance. Companies were allowed to deduct for depletion the higher of their depletion calculated on a cost basis or 27½ percent of revenues from the property. This percentage or statuatory depletion was taken regardless of the amount capitalized. In this way, over a period of years, companies were able to take substantially more than the original cost of the property in the form of depletion.

As the oil companies fell out of favor with Congress in the 1970's this depletion allowance was substantially eroded. Today the allowance is only 18 percent of revenues and will be reduced annually to 15 percent by 1984. Under current law it will stay at 15 percent.

The percentage or statutory depletion may be taken only by small independent oil companies. An independent producer is one which does not have substantial refining or retailing operations. In addition, the percentage allowance may be applied only to the first 1,000 barrels of production each day. The percentage depletion may also be applied to certain fixed price gas contracts.

In summary then, percentage depletion has been all but eliminated for large integrated oil companies, but is of real value to the small independent producer. For purposes of the following example, we have assumed that companies will calculate their tax depletion on the cost basis.

EXAMPLE

Refer to Figure 9-2. In this case the company owns two leases which are in different geographical locations. Both were acquired in the year of the example. Seismic studies were conducted on both with favorable results on lease A and unfavorable results on lease B. The engineers now determine lease B to be 100 percent impaired. A dry hole was drilled on lease A, followed by a second exploratory well which discovered proved reserves. We can assume that only one producing well is drilled. Notice that in each case the costs are divided into those which qualify as intangible and tangible costs. The working interest retains 42,000 of the 48,000 barrels produced during this tax year, with the royalty interest receiving the remainder. We will ignore the investment tax credit in this example. Under present law companies would be allowed to take a credit against their tax liability equal to 10 percent of the tangible costs which are capitalized.

INCOME STATEMENTS

Figure 9-3 contains the income statements from this example under successful efforts accounting, full cost accounting, and the tax basis. In each case the revenue is the same. The acquisition cost of lease B is deducted under both successful efforts and for the tax basis, because this property is deemed to be 100 percent impaired. It is capitalized under full costing. G&G costs in total are deducted under successful efforts. They are capitalized in total under full costing. Those G&G costs associated with lease A are capitalized for the tax basis accounting and those G&G costs associated with lease B are expensed for tax basis accounting. The exploratory dry hole is expensed for successful efforts accounting. All the intangible drilling and development costs from both wells, plus the intangible development of property A and the tangible costs on the abandoned property, are deducted for tax purposes. Lifting costs, severance taxes, and the windfall profit

TAX EXAMPLE

	LEASE A	LEASE B
Acquisition Costs	$ 10,000	$ 5,000
G and G Costs	$ 20,000	$15,000
G and G Results	Favorable	Unfavorable 100% Impaired
Exploratory Drilling (no.1)		
Costs: Tangible	$ 30,000	
Intangible	$300,000	
Results	Dry Hole	
Exploratory Drilling (no.2)		
Costs: Tangible	$ 40,000	
Intangible	$250,000	
Results	Proved Reserves	
Reserves Discovered	480,000 bbl	
Development Costs:		
Tangible	$300,000	
Intangible	$ 30,000	
Royalty Interest	1/8	
Sales Price	$42/bbl	
Lifting Costs	$5/bbl	
Severance Tax	10% of Revenue	
Windfall Profits Tax	$13/bbl	
Reserves Produced	48,000 bbl	
Tax Rate	46%	

Note: Ignore investment tax credit.

Figure 9–2

NET INCOME

	SE		FC		TAX BASIS	
Revenue		$1,764,000		$1,764,000		$1,764,000
Expenses:						
Acquisition	$ 5,000		–0–		$ 5,000	
G and G	35,000		–0–		15,000	
Exploratory Wells	330,000		–0–		580,000	
Development Costs	–0–		–0–		30,000	
Lifting Costs	240,000		240,000		240,000	
Severence Tax	176,400		176,400		176,400	
Windfall Profits Tax	546,000		546,000		546,000	
*Amortization	63,000	1,395,400	100,000	1,062,400	37,000	1,629,400
Net Income Before Income Taxes		$ 368,600		$ 701,600		$ 134,600
Provision for Income Taxes		169,556		322,736		61,916
Net Income		$ 199,044		$ 378,864		$ 72,684
*	$\frac{*\$630{,}000}{420{,}000 \text{ bbl}} \times 42{,}000 \text{ bbl}$		$\frac{*\$1{,}000{,}000}{420{,}000 \text{ bbl}} \times 42{,}000 \text{ bbl}$		$\frac{*\$370{,}000}{420{,}000 \text{ bbl}} \times 42{,}000 \text{ bbl}$	

Figure 9–3

tax are the same in all three methods. The amortization will be different for each method, in that the amount capitalized varies. Notice that the denominator is the same in each case. Provisions for income taxes are calculated as 46 percent of the net income before income taxes.

DEFERRED INCOME TAXES

The Accounting Principles Board created a dilemma for the accounting profession with APB Opinion No. 11. This opinion requires that a company show income tax expense (the provision for income taxes) as if the company had paid taxes on the accounting net income rather than according to the tax code. In our example the total tax due is equal to $74,336. But we must show an expense under successful efforts of $169,556, and under full costing of $322,736. The journal entries then create an imbalance; that is, if under successful efforts we debit the expense for $169,556 and credit the liability for $74,336 we do not have a journal entry that will balance.

Consequently, we must add an account called *Deferred Income Taxes* for the balance. These journal entries are illustrated in Figure 9-4. Deferred Income Taxes appears in the liability section of the balance sheet (if it has a credit balance, as in this case). An economic interpretation is that the company has deferred paying its taxes until later years; therefore, a liability or an obligation exists that the company will later have to meet.[1]

SUMMARY

We have shown that income tax accounting takes features of both full costing and successful efforts of the capitalization decision, and for small independent oil and gas producers will amortize on a basis unlike either of the financial accounting methods.

A final word of caution is in order. Many smaller companies keep their books on the tax basis because they have limited external reporting requirements outside of their annual tax return.

[1]Many accountants have taken issue with the advisability of recording deferred income taxes. They would prefer to show the provision for income taxes equal to the actual tax liability.

Therefore, their chart of accounts is set up on tangible versus intangible expenditures. The cost center is maintained on a property basis and, if they are eligible for percentage depletion, will not include detailed historical cost records. Substantial data base revisions would be needed then to prepare proper audited financial statements.

TAX PROVISION ENTRIES

SE		
Provision for Income Taxes	169,556	
Deferred Income Taxes		107,640
Taxes Payable		91,916
FC		
Provision for Income Taxes	322,736	
Deferred Income Taxes		260,820
Taxes Payable		61,916

Figure 9–4

Chapter 10

Other Financial Statement Disclosures

Our discussions up to this point have dealt with the two primary financial statements of oil and gas producing companies, the income statement and the balance sheet. A cursory review of the financial statements presented in Appendix B shows that a great deal more information is presented besides these two primary documents. In this chapter we will review the disclosures made by oil and gas producing companies that extend beyond the scope of the income statement and the balance sheet.

These disclosures fall generally within three categories: (1) disclosures required of any company representing its statements to be a complete set of audited financial statements; (2) additional disclosures required as part of the financial statements of companies reporting to the SEC, referred to as *Regulation S-X* disclosures; and (3) disclosures required by SEC registrants that are not part of the financial statements but still are required to be disclosed. These are referred to as *Regulation S-K* disclosures.

DISCLOSURES MADE BY ALL OIL AND GAS PRODUCING COMPANIES

Statement of Financial Accounting Standards No. 19 requires that three general categories of disclosures be included any time an oil and gas producing company represents a set of statements as being a complete set of financial statements. These are information on reserve quantity data, information on costs which have been capitalized, and information on costs which have been incurred.

Net quantities of proved oil and gas reserves must be disclosed for each year the company reports financial statements. These are divided into proved developed and total proved reserves. Of course, the astute reader can determine proved undeveloped reserves by merely subtracting the two numbers. The reserves refer to the reporting entity's mineral interest in the reserves; that is, if an oil company has a working interest equal to seven-eighths of the total interest of a property and that property contains 800,000 barrels of proved oil reserves, the entity would report 700,000 barrels as the amount of reserves to be shown in the financial statements. Notice that reserves are shown by geographical area and, in addition, quantities that were changed by purchases, sales acquisitions and revisions, also are disclosed. It should be pointed out here that the auditor does not express an opinion on the accuracy of this reserve information. For that reason these disclosures are clearly marked *unaudited*. However, the auditor's name is deemed to be associated with this information, so that he has given these numbers at least a cursory review. Certainly it is easy to understand that the certified public accountant usually is not an expert at petroleum reservoir engineering; therefore, it would be impossible for the auditor to express an opinion on this reserve data.

The second general disclosure to be made concerns capitalized costs. This includes the amount of gross costs capitalized for each of the general categories of assets of the oil and gas company. These capitalized costs are then shown net of accumulated amortization, or any valuation allowance. This information gives the statement reader some indication of the direction which the company's activities are taking. For example, a company which has a much larger investment in unproved properties than another company is likely to soon undertake a major exploration program.

The final category of disclosures which all companies must make concerns the costs incurred in particular activities. Specifically, the company must disclose how many dollars were spent in the last year in acquisition, exploration, development and production activities. This enables the statement reader to judge the amount of total activity the company is placing in each of these four activities. For example, a company that spends a relatively smaller amount on acquisition and exploration and a relatively larger amount on development and production is not likely to be adding proved reserves. The auditor does express an opinion on both the costs which have been capitalized and on the costs incurred.

FINANCIAL STATEMENT DISCLOSURES FOR SEC REGISTRANTS

These disclosures are referred to as Regulation S-X disclosures because they are required by that regulation which establishes accounting rules for the annual report filed by the companies with the SEC. Regulation S-X requires all of the disclosures previously mentioned, plus the items discussed below.

A very major disclosure is that which represents "value" of the proved reserves discovered and the proved reserves in place owned by the oil and gas producing company. This general set of disclosures is referred to as *Reserve Recognition Accounting*. See Appendix B for an illustration of these statements. Basically they refer to the economic benefit of the company's acquisition and exploration activities. That is, the statements attempt to measure the economic benefit of proved reserves added during the period and subtract the economic cost of finding those reserves. The difference is referred to as the economic results of oil and gas producing activities. These disclosures also describe how this net economic value changed during the year. A certain portion of the value was increased through revisions or additions to proved reserves. Other portions were reduced because the reserves were produced and sold. The second statement addresses these particular issues. All of these RRA numbers are computed using current prices and current costs. Future cash flows are then discounted back to present value at a discount rate of 10 percent.

In addition, the company must also disclose the revenues expected during the next three years individually and as a lump sum thereafter. There has been much controversy over the RRA and future revenue disclosures. The economic theory behind the statements certainly is valid, but the problems with RRA have centered on the lack of reliability of reserve estimates and the certain arbitrariness of the rules required by the SEC. Currently the SEC will continue to require supplemental unaudited disclosure of this data.

REGULATION S-K DISCLOSURES BY SEC REGISTRANTS

The final category of disclosures is referred to as Regulation S-K disclosures, because the disclosures are required by SEC Regulation S-K. The significant point here is that Regulation S-K disclosures are not subject to the auditor's report and are considered to be outside the financial statements. They do not need to be marked unaudited in order for the reader to know that the auditor has not performed standard tests on those items.

The first disclosure refers to the average sales price and the average production or lifting costs. This information is separated by geographic area to allow the financial statement reader to evaluate the potential revenues.

Companies must also disclose the gross and net wells reported by geographical area that are currently in production, as well as those wells drilled. Disclosure of wells drilled must be further broken down to development wells and exploratory wells which are drilled. These numbers must further be broken down to productive wells and dry holes. Gross wells refers to the total number of wells in which the company has any interest at all. Net wells refers to the percent of ownership of wells the company has. For example, if a company has a 20 percent ownership in five different wells, it reports ownership in five gross wells but only one net well. Finally, the companies must report the number of acres over which they hold leases. These are reported by developed and undeveloped acreage and also by geographical area. Again, the acres are shown both gross and net in the same manner as were the numbers reporting well ownership.

OTHER DISCLOSURES

There are many other disclosures which oil and gas producing companies must make in their financial statements, such as certain commitments with foreign governments, certain major discoveries, and certain projects which have been excluded from the amortization rate. These disclosures are outside the scope of this discussion; the interested reader is referred to ASR No. 257 in Appendix A and the examples in Appendix B.

SUMMARY

It is clear that the total package of disclosures by oil and gas producing companies has become enormous. One can easily question the desirability of having such a large number of numbers thrown at the financial statement user. For this and other reasons, the Financial Accounting Standards Board has undertaken a major project to review the entire disclosure package. As this Handbook goes to press the FASB is holding hearings to aid in determining which disclosures are relevant, reliable, and have an effective benefit relationship to the cost of preparing and disseminating information. In all probability the entire package will be consolidated under one set of rules and likely will be streamlined.

The SEC has indicated that it will review the disclosure requirements resulting from this FASB study and if those disclosure requirements are not inconsistent with SEC objectives, they also will be adopted by the SEC. Should this happen, we will then have one set of disclosure requirements for all companies.

It also is quite possible that the FASB will limit the disclosure package to only those large companies that meet certain size criteria.

In summary, we should observe that the entire body of information presented about the oil and gas producing company goes far beyond the bottom lines of the income statement and the balance sheet. The reported net income is an important number, but in order to truly evaluate the results of the operations of the company we must look at a large number of other factors. In the same light, the total assets reported on the balance sheet give us an indication of the size of the company, but in order to measure its true asset potential we must look at a great deal of facts beyond the historical costs of acquiring those assets. This has been the attempt of these disclosure requirements.

Appendix A

Current SEC Rules for Oil and Gas Producing Companies

ASR No. 257 (General Rules plus Successful Efforts Rules)

ASR No. 258 (Full Cost Rules)

SEC Successful Efforts Rules Accounting Series Release No. 257

[8010-01-M]

Title 17 — Commodity and Securities Exchanges

CHAPTER II — SECURITIES AND EXCHANGE COMMISSION

[Release Nos. 33-6006; 34-15416; 35-20836; IC-10530; AS-257]

PART 210 — FORM AND CONTENT OF FINANCIAL STATEMENTS, SECURITIES ACT OF 1933, SECURITIES EXCHANGE ACT OF 1934, PUBLIC UTILITY HOLDING COMPANY ACT OF 1935, INVESTMENT COMPANY ACT OF 1940, AND ENERGY POLICY AND CONSERVATION ACT OF 1975

Requirements for Financial Accounting and Reporting Practices for Oil and Gas Producing Activities

AGENCY: Securities and Exchange Commission.

ACTION: Amended rules.

SUMMARY: The Commission is adopting various amendments to its accounting and reporting requirements for oil and gas producers. These amendments: (1) Increase the conformity of rules relating to the successful efforts method of accounting to the provisions of Statement No. 19 of the Financial Accounting Standards Board; (2) revise certain definitions to correspond to those adopted by the Department of Energy for its Financial Reporting System; (3) provide an exemption from the disclosure requirements for diversified companies who meet specified criteria; and (4) clarify the application of the rules to cost-of-service regulated companies.

EFFECTIVE DATE: The rules published in this release shall be effective initially for fiscal years ending after December 25, 1978 that are contained in filings that include fiscal years ending after December 25, 1979. Certain portions are effective for purposes of the disclosure requirements of Regulation S-K for filings that include fiscal years ending after December 25, 1978.

FOR FURTHER INFORMATION CONTACT: James L. Russell, Office of the Chief Accountant, Securities and Exchange Commission, 500 North Capitol Street, Washington, D.C. 20549 (202-755-0222).

SUPPLEMENTARY INFORMATION: In Accounting Series Release No. 253 ("ASR No. 253"),[1] published on August 31, 1978, the Commission amended Regulation S-X, which governs the form and content of financial statements, to add new §210.3-18, establishing accounting and disclosure requirements for oil and gas producing companies. These rules apply to filings pursuant to the federal securities laws and to reports of energy information that are subject to the requirements of the Energy Policy and Conservation Act of 1975 ("EPCA"), ASR No. 253 specified the form of successful efforts accounting to be used by companies following

[1]Securities Act Release No. 33-5966 (43 FR 40688).

that method (a method identical to that contained in Statement of Financial Accounting Standards No. 19, "Financial Accounting and Reporting by Oil and Gas Producing Companies" ("FAS 19"), published in December 1977 by the Financial Accounting Standards Board).

In ASR No. 253, the Commission noted that the rules it was adopting were intended to meet certain objectives, including the following: (1) Conformity of the prescribed successful efforts method to the related provisions of FAS 19; (2) consistency with the Financial Reporting System under development by the Department of Energy ("DOE"); and (3) consistency of the revised definition of proved oil and gas reserves with other published definitions. The Commission requested comments as to whether these objectives had been met. In response to comments received on these aspects of the rules, the Commission has determined to make certain technical amendments to the adopted rules.

Further, the Commission has determined it appropriate to add certain exemptive provisions to the rules. These provisions exempt companies whose oil and gas producing activities do not exceed specified criteria from the disclosure requirements of §210.3-18(k) for purposes of filings under the federal securities laws. This release also clarifies the circumstances under which §210.3-18 may be deemed not to apply to rate-regulated companies.

Since these amendments are intended simply to dispense with technical inconsistencies in the adopted rules, or to provide limited exemptions from the rules under certain circumstances, and in light of the extensive public rule-making proceeding which led to the adoption of the rules, the Commission has determined that noticing these amendments for comment would be of little or no value. Therefore, the Commission for good cause finds that notice and public procedure on these amendments is unnecessary, as provided by section 553(b) of the Administrative Procedure Act (5 U.S.C. 553(b)).

Amendments to the Rules

A number of technical amendments to §210.3-18 are being made in response to requests that maximum conformity with the descriptive material and interpretive guidance of FAS 19 be achieved:

The definition of "exploratory well" in §210.3-18(a)(10) has been revised to include the wording used in FAS 19. In addition, definitions have been provided in §210.3-18(a) for "acquisitions of properties," "exploration costs," "development costs," and "production costs." These definitions incorporate into the rules a portion of the information that was stated in ASR No. 253 under "Description of the Rules." In particular, the definitions clarify the accounting for the different classifications of costs, and discuss the relationship of production of lifting costs (as well as depreciation, depletion and amortization) to the determination of the inventory cost of oil and gas.

Subparagraph (b)(5) has been added to specify the accounting for reimbursable costs of geological and geophysical studies, as contained in FAS 19.

Subparagraph (e)(3) has been revised to (1) adopt the wording of FAS 19 with respect to aggregation, for purposes of computing amortization, of large numbers of royalty interests whose acquisition costs are not individually significant; (2) include restoration costs among the costs to be considered in determining amortization and depreciation rates; and (3) permit amortization of joint oil and gas production to be based on the dominant mineral.

The phrase "as incurred" has been removed from the heading of paragraph (f). Also, dry hole and bottom hole contributions have been included among the examples of costs to be charged to expense.

Paragraph (g) has been amended to clarify that the accounting standards specifically prohibit "the deferral of costs of exploratory wells that find some oil and gas reserves merely on the chance that some event totally beyond the entity's control will occur."

Subparagraph (k)(5)(xii) now states specifically that the amount at which a receivable related to a retained production payment is recorded should include consideration of imputed interest.

Subparagraph (k)(2) has been modified to require disclosure of gross and net capitalized costs as of the end of each period for which "a complete set of (annual or interim) financial statements is presented," rather than as of the end of each period for which a balance sheet is presented. This will make it consistent with the corresponding requirements of FAS 19.

A discussion of initial retroactive application of the rules, similar to that contained in FAS 19, has been included in this release under "Effective Date."

Definition of Proved Reserves

ASR No. 253 adopted the definition of proved oil and gas reserves that was contained in DOE's Form EIA-23, "Annual Survey of Domestic Oil and Gas Reserves." A number of comment letters pointed out that this definition differs somewhat from that contained in DOE's Form EIA-28, '—'Energy Company Financial Reporting System" ("FRS"). Since the FRS is intended to collect information pursuant to EPCA for purposes of developing a national energy data base, most commentators urged that the definition of proved reserves adopted by the Commission conform to that contained in FRS.

Accordingly, the definitions of "proved oil and gas reserves," "proved developed oil and gas reserves," and "proved undeveloped reserves" set forth in §210.3-18(a) of Regulation S-X have been modified to correspond to those contained in Form EIA-28. Although there does not appear to be any substantive difference between the definitions, this conformity should eliminate any possible ambiguity concerning their use. However, the Commission's definition of proved oil and gas reserves continues to contain a discussion of the term "existing economic and operating conditions," to specify that current prices include consideration of changes in existing prices provided only by contractual arrangements.

Several commentators requested clarification as to whether the ' current prices" used in determining estimates of proved reserves (and also in computing estimated future net revenue pursuant to paragraph (k)(5) of §210.3-18) may incorporate price increases allowed by Federal Energy Regulatory Commission national and area rate opinions or by the Natural Gas Policy Act of 1978, if such price increases have been specifically incorporated into gas contracts. The Commission believes that consideration of such gas price escalations is appropriate only to the extent that amounts are fixed and determinable. Specifically, no recognition should be given to potential price increases related to uncertain events such as an "annual inflation factor."

Exemption From Disclosure Requirements

A number of comment letters included a request that an exemption be provided from the disclosure requirements of §210.3-18(k) for registrants whose oil and gas producing activities are not significant in relation to their total operations. The Commission has concluded that such an exemption is appropriate and has amended §210.3-18(k) to provide an exemption (for filings under the federal securities laws) from the disclosure requirements of that paragraph for registrants whose revenues, income and assets relating to oil and gas producing activities are less than 10 percent of the corresponding consolidated amounts for each of the two most recent fiscal years. For purposes of the asset criterion, proved oil and gas properties are to be measured on the basis of "the present value of estimated future net revenues," rather than historical cost. This exemption does not affect the general applicability of the financial accounting standards of §210.3-18.

Application of Rules to Regulated Companies

The rules for financial accounting and reporting practices of oil and gas producers adopted in ASR No. 253 did not specifically address the application of those requirements to regulated activities. The addendum to APB Opinion No. 2, "Accounting for the Investment Credit,"[2] states that "differences may arise in the application of generally accepted accounting principles as between regulated and nonregulated businesses, because of the effect in regulated businesses of the rate-making process," and discusses the application of generally accepted accounting principles to regulated businesses. The Commission recognizes that the oil and gas producing activities of some companies may be subject to regulatory requirements for rate-making purposes. Accordingly, the introduction to §210.3-18 has been amended to specify that, when oil and gas properties are regulated for rate-making purposes on an individual-company-cost-of-service basis, companies may give recognition to the effect of rate-making considerations

[2]This opinion was published in December 1962 by the Accounting Principles Board, which then existed as the standard-setting body of the American Institute of Certified Public Accountants.

on costs to be capitalized and on the basis for amortization. However, the Commission intends to reassess this matter upon completion of the Financial Accounting Standards Board's project involving reconsideration of the Addendum to APB Opinion No. 2.

Indexing

The indexing within paragraphs (a)(1), (k)(4), and (k)(5) has been revised to conform to that contained in the Federal Register's publication of ASR No. 253 (43 FR 40688).

Effective Date

The rules contained in §210.3-18 are effective for fiscal years ending after December 25, 1978, that are contained in filings that include fiscal years ending after December 25, 1979, although earlier application is encouraged. Certain portions of §210.3-18 are also effective for purposes of the disclosure requirements of Regulation S-K for filings that include fiscal years ending after December 25, 1978.

Accounting changes adopted to conform to the provisions of these rules are to be made retroactively by restating the financial statements of prior periods. Financial statements for the fiscal year in which these rules are first applied should disclose the nature of the accounting changes and their effect on income before extraordinary items, net income, and related per share amounts for each period restated.

Retroactive application of the provisions of §210.3-18 may require the use of estimates and approximations. A provision that would not have a significant effect on prior years' financial statements need not be retroactively applied. Further, retroactive application of some provisions of these rules may require the use of estimates of a type not previously made; information that may have become available some time after the year being restated may be taken into account in making those estimates, except that estimates of quantities of oil and gas reserves that had been made in prior years need not currently be revised in retrospect.

Commission Action

The Commissioner hereby amends 17 CFR Part 210 by revising §210.3-18. This section as amended is set forth below:

§210.3-18 Financial accounting and reporting for oil and gas producing activities pursuant to the Federal securities laws and the Energy Policy and Conservation Act of 1975.

This section prescribes financial accounting and reporting standards for registrants with the Commission engaged in oil and gas producing activities in filings under the Federal securities laws and for the preparation of accounts by persons engaged, in whole or in part, in the production of crude oil or natural gas in the United States, pursuant to section 503 of the Energy Policy and Conservation Act of 1975 (42 U.S.C. 6383) ("EPCA") and section 11(c) of the Energy Supply and Environmental Coordination Act of 1974 (15 U.S.C. 796) ("ESECA"), as amended by section 505 of EPCA. The application of this section to those oil and gas producing operations of companies regulated for rate-making purposes on an individual-company-cost-of-service basis may, however, give appropriate recognition to differences arising because of the effect of the rate-making process.

Exemption. Any person exempted by the Department of Energy from any record-keeping or reporting requirements pursuant to section 11(c) of ESECA, as amended, is similarly exempted from the related provisions of this section in the preparation of accounts pursuant to EPCA. This exemption does not affect the applicability of this section to filings pursuant to the Federal securities laws.

Definitions

(a) *Definitions.* The following definitions apply to the terms listed below as they are used in this section:

(1) *Oil and gas producing activities.* (i) Such activities include:

(A) The search for crude oil, including condensate and natural gas liquids, or natural gas ("oil and gas") in their natural states and original locations.

(B) The acquisition of property rights or properties for the purpose of further exploration and/or for the purpose of removing the oil or gas from existing reservoirs on those properties.

(C) The construction, drilling and production activities necessary to retrieve oil and gas from its natural reservoirs, and the acquisition, construction, installation, and maintenance of field gathering and storage systems — including lifting the oil and gas to the surface and gathering, treating, field processing (as in the case of processing gas to extract liquid hydrocarbons) and field storage. For purposes of this section, the oil and gas production function shall normally be regarded as terminating at the outlet valve on the lease or field storage tank; if unusual physical or operational circumstances exist, it may be appropriate to regard the production functions as terminating at the first point at which oil, gas, or gas liquids are delivered to a main pipeline, a common carrier, a refinery, or a marine terminal.

(ii) Oil and gas producing activities do not include:

(A) The transporting, refining and marketing of oil and gas.

(B) Activities relating to the production of natural resources other than oil and gas.

(C) The production of geothermal steam or the extraction of hydrocarbons as a by-product of the production of geothermal steam or associated geothermal resources as defined in the Geothermal Steam Act of 1970.

(D) The extraction of hydrocarbons from shale, tar sands, or coal.

(2) *Proved oil and gas reserves.* Proved oil and gas reserves are the estimated quantities of crude oil, natural gas, and natural gas liquids which geological and engineering data demonstrate with reasonable certainty to be recoverable in future years from known reservoirs under existing economic and operating conditions, i.e., prices and costs as of the date the estimate is made. Prices include consideration of changes in existing prices provided only by contractual arrangements, but not on escalations based upon future conditions.

(i) Reservoirs are considered proved if economic producibility is supported by either actual production or conclusive formation tests. The area of a reservoir considered proved includes (A) that portion delineated by drilling and defined by gas-oil and/or oil-water contacts, if any, and (B) the immediately adjoining portions not yet drilled, but which can be reasonably judged as economically productive on the basis of available geological and engineering data. In the absence of information on fluid contacts, the lowest known structural occurrence of hydrocarbons controls the lower proved limit of the reservoir.

(ii) Reserves which can be produced economically through application of improved recovery techniques (such as fluid injection) are included in the "proved" classification when successful testing by a pilot project, or the operation of an installed program in the reservoir, provides support for the engineering analysis on which the project or program was based.

(iii) Estimates of proved reserves do not include the following: (A) Oil that may become available from known reservoirs but is classified separately as "indicated additional reserves"; (B) crude oil, natural gas, and natural gas liquids, the recovery of which is subject to reasonable doubt because of uncertainty as to geology, reservoir characteristics, or economic factors; (C) crude oil, natural gas, and natural gas liquids, that may occur in undrilled prospects; and (D) crude oil, natural gas, and natural gas liquids, that may be recovered from oil shales, coal, gilsonite and other such sources.

(3) *Proved developed oil and gas reserves.* Proved developed oil and gas reserves are reserves that can be expected to be recovered through existing wells with existing equipment and operating methods. Additional oil and gas expected to be obtained through the application of fluid injection or other improved recovery techniques for supplementing the natural forces and mechanisms of primary recovery should be included as "proved developed reserves" only after testing by a pilot project or after the operation of an installed program has confirmed through production response that increased recovery will be achieved.

(4) *Proved undeveloped reserves.* Proved undeveloped oil and gas reserves are reserves that are expected to be recovered from new wells on undrilled acreage, or from existing wells where a relatively major expenditure is required for recompletion. Reserves on undrilled acreage shall be limited to those drilling units offsetting productive units that are reasonably certain of production when drilled. Proved reserves for other undrilled units can be claimed only where it can be demonstrated with certainty that there is continuity of production from the existing productive formation. Under no circumstances should estimates for proved undeveloped reserves be attributable to any acreage for which an

application of fluid injection or other improved recovery technique is contemplated, unless such techniques have been proved effective by actual tests in the area and in the same reservoir.

(5) *Proved properties.* Properties with proved reserves.

(6) *Unproved Properties.* Properties with no proved reserves.

(7) *Proved area.* The part of a property to which proved reserves have been specifically attributed.

(8) *Field.* An area consisting of a single reservoir or multiple reservoirs all grouped on or related to the same individual geological structural feature and/or stratigraphic condition. There may be two or more reservoirs in a field that are separated vertically by intervening impervious strata, or laterally by local geologic barriers, or by both. Reservoirs that are associated by being in overlapping or adjacent fields may be treated as a single or common operational field. The geological terms "structural feature" and "stratigraphic condition" are intended to identify localized geological features as opposed to the broader terms of basins, trends, provinces, plays, areas-of-interest, etc.

(9) *Reservoir.* A porous and permeable underground formation containing a natural accumulation of producible oil and/or gas that is confined by impermeable rock or water barriers and is individual and separate from other reservoirs.

(10) *Exploratory well.* A well drilled to find and produce oil or gas in an unproved area, to find a new reservoir in a field previously found to be productive of oil or gas in another reservoir, or to extend a known reservoir. Generally, an exploratory well is any well that is not a development well, a service well, or a stratigraphic test well as those items are defined below.

(11) *Development well.* A well drilled within the proved area of an oil or gas reservoir to the depth of a stratigraphic horizon known to be productive.

(12) *Service well.* A well drilled or completed for the purpose of supporting production in an existing field. Specific purposes of service wells include gas injection, water injection steam injection, air injection, saltwater disposal, water supply for injection, observation, or injection for insitu combustion.

(13) *Stratigraphic test well.* A drilling effort, geologically directed, to obtain information pertaining to a specific geologic condition. Such wells customarily are drilled without the intention of being completed for hydrocarbon production. This classification also includes tests identified as core tests and all types of ex-

pendable holes related to hydrocarbon exploration. Stratigraphic test wells are classified as (i) "exploratory-type," if not drilled in a proved area, or (ii) "development-type," if drilled in a proved area.

(14) *Acquisition of properties.* Costs incurred to purchase, lease or otherwise acquire a property, including costs of lease bonuses and options to purchase or lease properties, the portion of costs applicable to minerals when land including mineral rights is purchased in fee, brokers' fees, recording fees, legal costs, and other costs incurred in acquiring properties.

(15) *Exploration costs.* Costs incurred in identifying areas that may warrant examination and in examining specific areas that are considered to have prospects of containing oil and gas reserves, including costs of drilling exploratory wells and exploratory-type stratigraphic test wells. Exploration costs may be incurred both before acquiring the related property (sometimes referred to in part as prospecting costs) and after acquiring the property. Principal types of exploration costs, which include depreciation and applicable operating costs of support equipment and facilities and other costs of exploration activities, are:

(i) Costs of topographical, geographical and geophysical studies, rights of access to properties to conduct those studies, and salaries and other expenses of geologists, geophysical crews, and others conducting those studies. Collectively, these are sometimes referred to as geological and geophysical or "G&G" costs.

(ii) Costs of carrying and retaining undeveloped properties, such as delay rentals, ad valorem taxes on properties, legal costs for title defense, and the maintenance of land and lease records.

(iii) Dry hole contributions and bottom hole contributions.

(iv) Costs of drilling and equipping exploratory wells.

(v) Costs of drilling exploratory-type stratigraphic test wells.

(16) *Development costs.* Costs incurred to obtain access to proved reserves and to provide facilities for extracting, treating, gathering and storing the oil and gas. More specifically, development costs, including depreciation and applicable operating costs of support equipment and facilities and other costs of development activities, are costs incurred to:

(i) Gain access to and prepare well locations for drilling, including surveying well locations for the purpose of determining specific development drilling sites, clearing ground, draining, road building, and relocating public roads, gas lines, and power lines, to the extent necessary in developing the proved reserves.

(ii) Drill and equip development wells, development-type stratigraphic test wells, and service wells, including the costs of platforms and of well equipment such as casing, tubing, pumping equipment, and the wellhead assembly.

(iii) Acquire, construct, and install production facilities such as lease flow lines, separators, treaters, heaters, manifolds, measuring devices, and production storage tanks, natural gas cycling and processing plants, and central utility and waste disposal systems.

(iv) Provide improved recovery systems.

(17) *Production costs.* (i) Costs incurred to operate and maintain wells and related equipment and facilities, including depreciation and applicable operating costs of support equipment and facilities and other costs of operating and maintaining those wells and related equipment and facilities. They become part of the cost of oil and gas produced. Examples of production costs (sometimes called lifting costs) are:

(A) Costs of labor to operate the wells and related equipment and facilities.

(B) Repairs and maintenance.

(C) Materials, supplies, and fuel consumed and supplies utilized in operating the wells and related equipment and facilities.

(D) Property taxes and insurance applicable to proved properties and wells and related equipment and facilities.

(E) Severance taxes.

(ii) Some support equipment or facilities may serve two or more oil and gas producing activities and may also serve transportation, refining, and marketing activities. To the extent that the support equipment and facilities are used in oil and gas producing activities, their depreciation and applicable operating costs become exploration, development or production costs, as appropriate. Depreciation, depletion, and amortization of capitalized acquisition, exploration, and development costs are not production costs but also become part of the cost of oil and gas produced along with production (lifting) costs identified above.

SUCCESSFUL EFFORTS METHOD

(b) *Costs to be capitalized if the successful efforts method of accounting is followed.* The costs of the following assets involved in oil and gas producing activities are to be capitalized when incurred:

(1) *Mineral interests in properties.* Including (i) fee ownership or a lease, concession, or other interest representing the right to extract oil or gas subject to such terms as may be imposed by the conveyance of that interest; (ii) royalty interests, production payments payable in oil or gas, and other nonoperating interests in properties operated by others; and (iii) those agreements with foreign governments or authorities under which a reporting entity participates in the operation of the related properties or otherwise serves as "producer" of the underlying reserves (as opposed to being an independent purchaser, broker, dealer, or importer). Properties do not include other supply agreements or contracts that represent the right to purchase, rather than extract, oil and gas.

(2) *Wells and related equipment and facilities.* Including: (i) Costs incurred to drill and equip those exploratory wells and exploratory-type stratigraphic test wells that have found proved reserves and (ii) development costs, i.e., costs incurred to obtain access to proved reserves and provide facilities for extracting, treating, gathering, and storing the oil and gas, including the drilling and equipping of development wells and development-type stratigraphic test wells (whether those wells are successful or unsuccessful) and service wells.

(3) *Support equipment and facilities used in oil and gas producing activities.* Items such as seismic equipment, drilling equipment, construction and grading equipment, vehicles, repair shops, warehouses, supply points, camps, and division, district, or field offices.

(4) *Uncompleted wells, equipment and facilities.* Including costs incurred to: (i) Drill and equip wells that are not yet completed and (ii) acquire or construct equipment and facilities that are not yet completed and installed.

(5) *Geological and geophysical studies.* G&G studies may be conducted on a property owned by another person, in exchange for an interest in the property if proved reserves are found or to be reimbursed if proved reserves are not found. In such cases, the G&G costs shall be accounted for as a receivable when incurred

and, if proved reserves are found, they shall become the cost of the proved property acquired.

(c) *Assessment of unproved properties if the successful efforts method of accounting is followed.* Unproved properties shall be assessed periodically to determine whether they have been impaired. A property would likely be impaired, for example, if a dry hole has been drilled on it and the reporting entity has no firm plans to continue drilling. Also, the likelihood of partial or total impairment of a property increases as the expiration of the lease term approaches if drilling activity has not commenced on the property or on nearby properties. Information that becomes available after the end of the period covered by the financial statements but before those financial statements are issued shall be taken into account in evaluating conditions that existed at the balance sheet date.

(1) If the results of the assessment indicate impairment, a loss shall be recognized by providing a valuation allowance. Impairment of individual unproved properties whose acquisition costs are relatively significant shall be assessed on a property-by-property basis, and an indicated loss shall be recognized by providing a valuation allowance. When a reporting entity has a relatively large number of unproved properties whose acquisition costs are not individually significant, it may not be practical to assess impairment on a property-by-property basis, in which case the amount of loss to be recognized and the amount of the valuation allowance needed to provide for impairment of those properties shall be determined by amortizing those properties, either in the aggregate or by groups, on the basis of the experience of the entity in similar situations and other information about such factors as the primary lease terms of those properties, the average holding period of unproved properties, and the relative proportion of such properties on which proved reserves have been found in the past.

(2) A property shall be reclassified from unproved properties to proved properties when proved reserves are discovered on or otherwise attributed to the property. Occasionally, a single property such as a foreign lease or concession covers so vast an area that only the portion of the property to which the proved reserves relate — determined on the basis of geological structural features or stratigraphic conditions — should be reclassified from unproved to proved. For a property whose impairment has been assessed

individually, the net carrying amount (acquisition cost minus valuation allowance) shall be reclassified to proved properties; for properties amortized by providing a valuation allowance on a group basis, the gross acquisition cost shall be reclassified.

(d) *Surrender or abandonment of properties if the successful efforts method of accounting is followed.* When an unproved property is surrendered, abandoned, or otherwise deemed worthless, capitalized acquisition costs relating thereto shall be charged against the related allowance for impairment to the extent an allowance has been provided; if the allowance previously provided is inadequate, a loss shall be recognized. Normally, no gain or loss shall be recognized if only an individual well or individual item of equipment is abandoned or retired or if only a single lease or other part of a group of proved properties constituting the amortization base is abandoned or retired as long as the remainder of the property or group of properties continues to produce oil or gas. Instead, the asset being abandoned or retired shall be deemed to be fully amortized, and its costs shall be charged to accumulated depreciation, depletion, or amortization. When the last well on an individual property (if that is the amortization base) or group of properties (if amortization is determined on the basis of an aggregation of properties with a common geological structure) ceases to produce and the entire property or property group is abandoned, gain or loss shall be recognized. Occasionally, the partial abandonment or retirement of a proved property or group of proved properties or the abandonment or retirement of wells or related equipment or facilities may result from a catastrophic event or other major abnormality. In those cases, a loss shall be recognized at the time of abandonment or retirement.

(e) *Amortization of capitalized costs if the successful effects method of accounting is followed.* (1) Capitalized acquisition costs of proved properties shall be amortized using the unit-of-production method on the basis of total estimated units of proved oil and gas reserves.

(2) Capitalized costs of exploratory wells and exploratory-type stratigraphic test wells that have found proved reserves and capitalized development costs shall be amortized (depreciated) using the unit-of-production method on the basis of total estimated units of proved developed reserves. However, it may be more appropriate, in some cases, to depreciate natural gas cy-

cling and processing plants by a method other than the unit-of-production method.

(3) Amortization, using the unit-of-production method as required by subparagraphs (1) and (2) of this paragraph, may be computed either on a property-by-property basis or on the basis of some reasonable aggregation of properties with a common geological structural feature or stratigraphic condition, such as a reservoir or field. When a reporting entity has a relatively large number of royalty interests whose acquisition costs are not individually significant, they may be aggregated, for purpose of computing amortization, without regard to commonality of geological structural features or stratigraphic conditions; if information is not available to estimate reserve quantities applicable to royalty interests owned, a method other than the unit-of-production method may be used to amortize their acquisition costs. If significant development costs (such as the cost of an offshore production platform) are incurred in connection with a planned group of development wells before all of the planned wells have been drilled, it will be necessary to exclude a portion of those development costs in determining the unit-of-production amortization rate until the additional development wells are drilled. Similarly, it will be necessary to exclude, in computing the amortization rate, those proved developed reserves that will be produced only after significant additional development costs are incurred, such as for improved recovery systems. However, in no case should future development costs be anticipated in computing the amortization rate. Estimated dismantlement, restoration, and abandonment costs and estimated residual salvage values shall be taken into account in determining amortization and depreciation rates. For those properties or groups of properties containing both oil reserves and gas reserves, the units of oil and gas used to compute amortization shall be converted to a common unit of measure on the basis of their approximate relative energy content (without considering their relative sales values). However, if the relative proportion of gas and oil extracted in the current period is expected to continue throughout the remaining productive life of the property, unit-of-production amortization may be computed on the basis of one of the two minerals only; similarly, if either oil or gas clearly dominates both the reserves and the current production (with dominance determined on the basis of relative energy content), unit-of-production amortization may be computed on the

basis of the dominant mineral only. Unit-of-production amortization rates shall be revised whenever there is an indication of the need for revision, but at least once a year; those revisions shall be accounted for prospectively as changes in accounting estimates.

(f) *Costs to be charged to expense if the successful efforts method of accounting is followed.* Costs incurred in oil and gas producing activities other than those described in paragraph (b) of this section shall be charged to expense. Examples include geological and geophysical costs, costs of carrying and retaining undeveloped properties, dry hole and bottom hole contributions, the costs of drilling those exploratory wells and exploratory-type stratigraphic test wells that do not find proved reserves (see paragraph (g) of this section), and the cost of oil and gas produced.

(g) *Accounting for the costs of exploratory wells and exploratory-type stratigraphic test wells if the successful efforts method of accounting is followed.* The costs of drilling exploratory wells and the costs of drilling exploratory-type stratigraphic test wells shall be capitalized as part of the reporting entity's uncompleted wells, equipment, and facilities pending determination of whether the well has found proved reserves. If the well has found proved reserves, the capitalized costs of drilling the well shall become part of the entity's wells and related equipment and facilities (even though the well may not be completed as a producing well); if, however, the well has not found proved reserves, the capitalized costs of drilling the well, net of any salvage value, shall be charged to expense. The determination of whether proved reserves are found is usually made on or shortly after completion of drilling the well, and the capitalized costs shall either be charged to expense or be reclassified as part of the costs of wells and related equipment and facilities at that time. Information that becomes available after the end of the period covered by the financial statements but before those financial statements are issued shall be taken into account in evaluating conditions that existed at the balance sheet date. Occasionally, an exploratory well or an exploratory-type stratigraphic test well may be determined to have found oil and gas reserves, but classification of those reserves as proved cannot be made when drilling is completed. In those cases, one of three subparagraphs set forth below shall apply. Subparagraphs (1) and (2) are intended to prohibit, in all cases, the deferral of the costs of exploratory wells that find some oil and gas reserves merely on the chance that some event totally beyond the entity's control will occur, e.g., on the chance that the selling prices of oil and gas will increase sufficiently to result in classification of reserves as proved that are not commercially recoverable at current prices.

(1) Exploratory wells that find oil and gas reserves in an area requiring a major capital expenditure, such as a trunk pipeline, before production could begin. On completion of drilling, an exploratory well may be determined to have found oil and gas reserves, but classification of those reserves as proved depends on whether a major capital expenditure can be justified which, in turn, depends on whether additional exploratory wells find a sufficient quantity of additional reserves. In that case, the cost of drilling the exploratory well shall continue to be carried as an asset pending determination of whether proved reserves have been found only as long as both of the following conditions are met: (i) The well has found a sufficient quantity of reserves to justify its completion as a producing well if the required capital expenditure is made, and (ii) drilling of the additional exploratory wells is under way or firmly planned for the near future. Otherwise, the exploratory well shall be assumed to be impaired, and its costs shall be charged to expense.

(2) All other exploratory wells that find oil and gas reserves. In the absence of a determination as to whether the reserves that have been found can be classified as proved, the costs of drilling such an exploratory well shall not be carried as an asset for more than one year following completion of drilling. If, after that year has passed, a determination that proved reserves have been found cannot be made, the well shall be assumed to be impaired, and its costs shall be charged to expense.

(3) Exploratory-type stratigraphic test wells that find oil and gas reserves. On completion of drilling, such a well may be determined to have found oil and gas reserves, but classification of those reserves as proved depends on whether a major capital expenditure (usually a production platform) can be justified which, in turn, depends on whether additional exploratory-type stratigraphic test wells find a sufficient quantity of additional reserves. In that case, the cost of drilling the exploratory-type stratigraphic test well shall continue to be carried as an asset pending determination of whether proved reserves have been found only as long as both of the following conditions are met: (i) The

well has found a quantity of reserves that would justify its completion for production had it not been simply a stratigraphic test well, and (ii) drilling of the additional exploratory-type stratigraphic test wells is under way or firmly planned for the near future. Otherwise, the exploratory-type stratigraphic test well shall be assumed to be impaired, and its costs shall be charged to expense.

(h) *Mineral property conveyances and related transactions if the successful efforts method of accounting is followed.* (1) Certain transactions, sometimes referred to as conveyances, are in substance borrowings repayable in cash or its equivalent and shall be accounted for as borrowings. The following are examples of such transactions:

(i) Entities seeking supplies of oil or gas sometimes make cash advances to operators to finance exploration in return for the right to purchase oil or gas discovered. Funds advanced for exploration that are repayable by offset against purchases of oil or gas discovered, or in cash if insufficient oil or gas is produced by a specified date, shall be accounted for as a receivable by the lender and as a payable by the operator.

(ii) Funds advanced to an operator that are repayable in cash out of the proceeds from a specified share of future production of a producing property, until the amount advanced plus interest at a specified or determinable rate is paid in full, shall be accounted for as a borrowing. The advance is a payable for the recipient of the cash and a receivable for the party making the advance. Such transactions, as well as those described in subparagraph (5)(i) of this paragraph, are commonly referred to as production payments. The two types differ in substance, however, as explained in subparagraph (5)(i) of this paragraph.

(2) In the following types of conveyances, gain or loss shall not be recognized at the time of conveyance:

(i) A transfer of assets used in oil and gas producing activities (including both proved and unproved properties) in exchange for other assets also used in oil and gas producing activities.

(ii) A pooling of assets in a joint undertaking intended to find, develop, or produce oil or gas from a particular property or group of properties.

(3) In the following types of conveyances, gain shall not be recognized at the time of the conveyance:

(i) A part of an interest owned is sold and substantial uncertainty exists about recovery of the costs applicable to the retained interest.

(ii) A part of an interest owned is sold and the seller has a substantial obligation for future performance, such as an obligation to drill a well or to operate the property without proportional reimbursement for that portion of the drilling or operating costs applicable to the interest sold.

(4) If a conveyance is not one of the types described in subparagraphs (2) and (3) of this paragraph, gain or loss shall be recognized unless there are other aspects of the transaction that would prohibit such recognition under accounting principles applicable to enterprises in general.

(5) In accordance with subparagraphs (2) through (4) of this paragraph, the following types of transactions shall be accounted for as indicated in each example. No attempt has been made to include the many variations of those arrangements that occur, but subparagraphs (2) through (4) of this paragraph shall, where applicable, determine the accounting for those other arrangements as well.

(i) Some production payments differ from those described in subparagraph (1)(ii) of this paragraph in that the seller's obligation is not expressed in monetary terms but as an obligation to deliver, free and clear of all expenses associated with operation of the property, a specified quantity of oil or gas to the purchaser out of a specified share of future production. Such a transaction is a sale of a mineral interest for which gain shall not be recognized because the seller has a substantial obligation for future performance. The seller shall account for the funds received as unearned revenue to be recognized as the oil or gas is delivered. The purchaser of such a production payment has acquired an interest in a mineral property that shall be recorded at cost and amortized by the unit-of-production method as delivery takes place. The estimated oil or gas reserves and production data shall be reported, in accordance with paragraph (k) of this section, as those of the purchaser of the production payment and not of the seller.

(ii) An assignment of the operating interest in an unproved property with retention of a nonoperating interest in return for drilling, development and operation by the assignee is a pooling of assets in a joint undertaking for which the assignor shall not recognize gain or loss. The assignor's cost of the original interest shall become the cost of the interest retained. The assignee shall account for all costs incurred as specified by paragraphs (b) through (g) of this section and shall allocate none of those costs to the mineral interest acquired. If oil or gas is discovered, each party shall report its share of oil and gas reserves and production, in accordance with paragraph (k) of this section.

(iii) An assignment of a part of an operating interest in an unproved property in exchange for a "free well" with provision for joint ownership and operation is a pooling of assets in a joint undertaking by the parties. The assignor shall record no cost for the obligatory well; the assignee shall record no cost for the mineral interest acquired. All drilling, development, and operating costs incurred by either party shall be accounted for as provided in paragraphs (b) through (g) of this section. If the conveyance agreement requires the assignee to incur geological or geophysical expenditures instead of, or in addition to, a drilling obligation, those costs shall likewise be accounted for by the assignee as provided in paragraphs (b) through (g) of this section. If reserves are discovered, each party shall report its share of reserves and production, in accordance with paragraph (k) of this section.

(iv) A part of an operating interest in an unproved property may be assigned to effect an arrangement called a "carried interest" whereby the assignee (the carrying party) agrees to defray all costs of drilling, developing, and operating the property and is entitled to all of the revenue from production from the property, excluding any third party interest, until all of the assignee's costs have been recovered, after which the assignor will share in both costs and production. Such an arrangement represents a pooling of assets in a joint undertaking by the assignor and assignee. The carried party shall make no accounting for any costs and revenue until after recoupment (payout) of the carried costs by the carrying party. Subsequent to payout the carried party shall account for its share of revenue, operating expenses, and (if the agreement provides for subsequent sharing of costs rather than a carried interest) subsequent development costs. During the payout period the carrying party shall record all costs, including those carried, as provided in paragraphs (b) through (g) of this section, and shall record all revenue from the property including that applicable to the recovery of cost carried. The carried party shall report as oil or gas reserves, in accordance with paragraph (k) of this section, only its share of proved reserves estimated to remain after payout, and unit-of-production amortization of the carried party's property cost shall not commence prior to payout. Prior to payout the carrying party's reserve estimates and production data, reported in accordance with paragraph (k) of this section, shall include the quantities applicable to recoupment of the carried costs.

(v) A part of an operating interest owned may be exchanged for a part of an operating interest owned by another party. The purpose of such an arrangement, commonly called a joint venture in the oil and gas industry, often is to avoid duplication of facilities, diversity risks, and achieve operating efficiencies. Such reciprocal conveyances represent exchanges of similar productive assets, and no gain or loss shall be recognized by either party at the time of the transaction. In some joint ventures which may or may not involve an exchange of interests, the parties may share different elements of costs in different proportions. In such an arrangement a party may acquire an interest in a property or in wells and related equipment that is disproportionate to the share of costs bourne by it. As in the case of a carried interest or a free well, each party shall account for its own cost under the provisions of this section. No gain shall be recognized for the acquisition of an interest in joint assets, the cost of which may have been paid in whole or in part by another party.

(vi) In a unitization all of the operating and nonoperating participants pool their assets in a producing area (normally a field) to form a single unit and in return receive an undivided interest (of the same type as previously held) in that unit. Unitizations generally are undertaken to obtain operating efficiencies and to enhance recovery of reserves, often through improved recovery operations. Participation in the unit is generally proportionate to the oil and gas reserves contributed by cash. Because the properties may be in different stages of development at the time of unitization, some participants may pay cash and others may receive cash to equalize contributions of wells and related equipment and facilities with the ownership interests in reserves. In those circumstances, cash paid by a participant shall be recorded as an additional investment in wells and related equipment and facilities, and cash received by a participant shall be recorded as a recovery of cost. The cost of the assets contributed plus or minus cash paid or received is the cost of the participant's undivided interest in the assets of the unit. Each participant shall include its interest in reporting reserve estimates and production data.

(vii) If the entire interest in an unproved property is sold for cash or cash equivalent, recognition of gain or loss depends on whether, in applying paragraph (c) of this section, impairment had been assessed for that property individually, or by amortizing that property as part of a group. If impairment was assessed individually, gain or loss shall be recognized. For a property amortized by providing a valuation allowance on a group basis, neither gain nor loss shall be recognized when an unproved property is sold unless the sales price exceeds the original cost of the property, in which case gain shall be recognized in the amount of such excess.

(viii) If a part of the interest in an unproved property is sold, even though for cash or cash equivalent, substantial uncertainty usually exists as to recovery of the cost applicable to the interest retained. Consequently, the amount received shall be treated as a recovery of cost. However, if the sales price exceeds the carrying amount of a property whose impairment has been assessed individually in accordance with paragraph (c) of this section, or exceeds the original cost of a property amortized by providing a valuation allowance on a group basis, gain shall be recognized in the amount of such excess.

(ix) The sale of an entire interest in a proved property that constitutes a separate amortization base is not one of the types of conveyances described in paragraphs (h)(2) and (h)(3) of this section. The difference between the amount of sales proceeds and the unamortized cost shall be recognized as a gain or loss.

(x) The sale of a part of a proved property, or of an entire proved property constituting a part of an amortization base, shall be accounted for as the sale of an asset, and a gain or loss shall be recognized, since it is not one of the conveyances described in paragraphs (h)(2) and (h)(3) of this section. The unamortized cost of the property or group of properties a part of which was sold shall be apportioned to the interest sold and the interest retained on the basis of the fair values of those interests. However, the sale may be accounted for as a normal retirement under the provisions of paragraph (d) of this section with no gain or loss recognized if doing so does not significantly affect the unit-of-production amortization rate.

(xi) The sale of the operating interest in a proved property for cash with retention of a nonoperating interest is not one of the types of conveyances described in paragraphs (h)(2) and (h)(3) of this section. Accordingly, it shall be accounted for as the sale of an asset, and any gain or loss shall be recognized. The seller shall allocate the cost of the proved property to the operating interest sold and the nonoperating interest retained on the basis of the fair values of those interests.

(xii) The sale of a proved property subject to a retained production payment that is expressed as a fixed sum of money payable only from a specified share of production from that property, with the purchase of the property obligated to incur the future costs of operating the property, shall be accounted for as follows: If satisfaction of the retained production payment is reasonably assured, the seller of the property, who retained the production payment, shall record the transaction as a sale, with recognition of any resulting gain or loss. The retained production payment shall be recorded as a receivable, at an amount reflecting appropriate consideration of imputed interest thereon. The purchaser shall record as the cost of the assets acquired the cash consideration paid plus the present value of the retained production payment, which shall be recorded as a payable. The oil and gas reserve estimates and production data, including those applicable to liquidation of the retained production payment, shall be reported by the purchaser of the property. If satisfaction of the retained production payment is not reasonably assured, the transaction is in substance a sale with retention of an overriding royalty that shall be accounted for in accordance with subparagraph (5)(xi) of this paragraph.

(xiii) The sale of a proved property subject to a retained production payment that is expressed as a right to a specified quantity of oil or gas out of a specified share of future production shall be accounted for in accordance with subparagraph (5)(xi) of this paragraph.

FULL COST METHOD

*   *   *   *   *

(Provisions of paragraph (i) are contained in Accounting Series Release No. 258 dated December 19, 1978.)

*   *   *   *   *

Income Taxes

(j) *Income taxes.* Comprehensive interperiod income tax allocation by the deferred method shall be followed for intangible drilling and development costs and other costs incurred that enter into the determination of taxable income and pretax accounting income in different periods. The excess of statutory depletion over cost depletion for income tax purposes shall be accounted for as a permanent difference in the period in which the excess is deducted for income tax purposes.

Disclosure Requirements

(k) *Disclosure of quantities of proved oil and gas reserves and historical financial data.* The following data shall be disclosed in the body of the financial statements, in the notes thereto, or in a separate schedule or other presentation that is an integral part of the financial statements.

Exemption. This paragraph shall not apply to filings under the federal securities laws by any registrant meeting all of the conditions described below for each of the two most recent fiscal years, based on its annual consolidated financial statements:

(i) Gross revenues from sales or transfers of oil and gas (as defined in subparagraph (4)(ii) of this paragraph) do not exceed 10 percent of total revenues;

(ii) Income after taxes (but before extraordinary items) from oil and gas producing activities, including amounts applicable to investees accounted for under the equity method, does not exceed 10 percent of consolidated income before extraordinary items; and

(iii) The "Present Value of Estimated Future Net Revenues" (see subparagraph (6)(ii) of this paragraph), including amounts attributable to investees accounted for under the equity method, plus the net capitalized costs of unproved properties, do not exceed 10 percent of total assets.

(1) *Disclosure of method of accounting.* The reporting entity shall disclose on the face of its balance sheet that it is adhering to the successful efforts method of accounting or the full cost method of accounting.

(2) *Disclosure of capitalized costs.* The aggregate amount of capitalized costs relating to oil and gas producing activities and

the aggregate amount of the related accumulated depreciation, depletion, amortization, and valuation allowances shall be reported as of the end of each period for which a complete set of (annual or interim) financial statements is presented, with separate presentation for each geographic area for which quantities of proved reserves are presented in accordance with paragraph (k)(5) of this section. If the capitalized costs of unproved properties are significant, aggregate amounts shall be reported separately for capitalized costs related to unproved properties and capitalized costs related to proved properties. Capitalized costs of support equipment and facilities may be disclosed separately or included, as appropriate, with capitalized costs of proved and unproved properties.

(3) *Disclosure of costs incurred in oil and gas producing activities.* The financial statements shall disclose the amounts of each of the following types of costs for each year for which an income statement is required (whether those costs are capitalized or charged to expense at the time they are incurred):

(i) Property acquisition costs (disclose separately the costs of acquiring proved properties, if significant).

(ii) Exploration costs.

(iii) Development costs.

(iv) Production (lifting) costs.

Exploration, development, and production costs include depreciation of support equipment and facilities used in those activities rather than the expenditures to acquire support equipment and facilities. Production (lifting) costs do not include depreciation, depletion, and amortization of capitalized acquisition, exploration, and development costs. If some or all of those costs are incurred in foreign countries, the amounts shall be disclosed separately for each of the geographic areas for which reserve quantities are disclosed in accordance with paragraph (k)(5) of this section and for any other foreign geographic area in which significant costs have been incurred without discovery of significant proved reserves. Also, disclose the aggregate amount of depreciation, depletion, amortization, and valuation provisions relating to oil and gas producing activities, presented separately for each of these geographic areas. If the reporting entity's share of the oil and gas reserves of its investee companies accounted for by the equity method is reported in conformity with paragraph (k)(5) of this section, disclosure shall also be made of the investor's share of its investee's capitalized costs and costs incurred in oil and gas producing activities.

(4) *Revenues from producing oil and gas.* (i) For each full fiscal year for which an income statement is required, disclose, for each geographic area for which reserve quantities are disclosed in accordance with paragraph (k)(5) of this section, the net revenues from oil and gas production related to each of the following, if significant: (A) Proved developed oil and gas reserves, (B) reserves applicable to long-term supply or similar agreements with foreign governments in which the entity acts as producer (see paragraph (k)(5) of this section), and (C) the entity's proportional interest in reserves of investees accounted for by the equity method.

(ii) Net revenues shall be computed by subtracting production (lifting) costs from gross revenues that are determined in accordance with the provisions of this subparagraph, as presented below. The following accounting practices shall be applied in determining gross revenues from sales or transfers of production of oil and gas:

(A) Sales to unaffiliated entities shall be included in gross revenues at the amount received in sales transactions attributable to net working interests, royalty interests, oil payments interests (see paragraph (h) of this section), net profits interests, etc., of the reporting entity. Production or severance taxes should not be deducted in determining gross revenues. Royalty payments and net profits disbursements should be excluded from gross revenues.

(B) Sales and transfers to unconsolidated affiliated persons and to other operations (such as refineries, chemical plants, etc.) of the reporting entity shall be separately disclosed and accounted for in the same manner as described in subparagraph (4)(ii)(A) of this paragraph except that such sales and transfers shall be valued at estimated market prices based on the prices of comparable products using posted field prices, if applicable, or amounts estimated to represent prices equivalent to those that could be obtained in a competitive arm's-length market environment, giving recognition to transportation costs, quality differences, and arrangements with and regulation by governments.

(5) *Disclosure of estimated quantities of proved oil and gas reserves.* Net quantities of proved reserves and proved developed reserves of crude oil (including condensate and natural gas li-

quids) and natural gas shall be reported as of the beginning and the end of each fiscal year for which an income statement is required. "Net" quantities of reserves include those relating to the operating and non-operating interests in properties as defined in paragraph (b)(1) of this section. Quantities of reserves relating to royalty interests owned shall be included in "net" quantities if the necessary information is available; if reserves relating to royalty interests owned are not included because the information is unavailable, that fact and the reporting entity's share of oil and gas produced for those royalty interests shall be reported for each year for which a complete set of financial statements is presented. "Net" quantities shall not include reserves relating to interests of others in properties owned by the reporting entity.

(i) Changes in the net quantities of proved reserves of oil and of gas during each fiscal year for which an income statement is required shall be reported. Changes resulting from each of the following shall be shown separately, with appropriate explanation of significant changes:

(A) *Revisions of previous estimates.* Revisions represent changes in previous estimates of proved reserves, either upward or downward, resulting from new information (except for an increase in proved acreage) normally obtained from development drilling and production history or resulting from a change in economic factors.

(B) *Improved recovery.* Changes in reserve estimates resulting from application of improved recovery techniques shall be separately shown if significant. If not significant, such changes shall be included in revisions of previous estimates.

(C) *Purchases of minerals-in-place.*

(D) *Extensions, discoveries, and other additions.* Additions to proved reserves that result from extension of the proved acreage of previously discovered (old) reservoirs through additional drilling in periods subsequent to discovery and discovery of new fields with proved reserves or of new reservoirs of proved reserves in old fields.

(E) *Production.*

(F) *Sales of minerals-in-place.*

(ii) If the reporting entity's proved reserves of oil and gas are located entirely within its home country, that fact shall be disclosed. If some or all of its reserves are located in foreign countries, the disclosures of net quantities of reserves of oil and gas and changes in them (as required above) shall be separately reported for the entity's home country (if significant reserves are located there) and each foreign geographic area in which significant reserves are located. Foreign geographic areas are individual countries or groups of countries, as appropriate, for meaningful disclosure in the circumstances.

(iii) Net quantities disclosed shall not include oil or gas subject to purchase under long-term supply, purchase, or similar agreements and contracts, including such agreements with foreign governments or authorities. However, quantities of oil and gas subject to such agreements with foreign governments or authorities as of the end of each fiscal year for which an income statement is required, and the net quantity of oil or gas received under the agreements during each such year, shall be separately disclosed if the reporting entity participates in the operation of the properties in which the oil or gas is located or otherwise serves as the "producer" of those reserves, as opposed, for example, to being an independent purchaser, broker, dealer, or importer.

(iv) In determining the reserve quantities to be reported:

(A) If consolidated financial statements are issued, 100 percent of the net reserve quantities attributable to the parent company and 100 percent of the net reserve quantities attributable to its consolidated subsidiaries (whether or not wholly owned) shall be included.

(B) If the financial statements include investments that are proportionately consolidated, reserve quantities shall include a proportionate share of the investee's net oil and gas reserves.

(C) If the financial statements include investments that are accounted for by the equity method, the investee's net proved oil and gas reserves shall not be included in the disclosures of reserves. However, the reporting entity's share of the investee's net proved oil and gas reserves shall be separately reported as of the end of each fiscal year for which an income statement is required.

(v) If important economic factors or significant uncertainties affect particular components of proved reserves, explanations shall be provided. Examples include unusually high expected development or lifting costs; the necessity to build a major pipeline or other major facilities before production of the reserves can begin; or contractual obligations to produce and sell a significant portion of reserves at prices that are substantially below those at which the oil and gas could otherwise be sold in the absence of the contractual obligation.

(vi) In reporting reserve quanitites and changes in them, oil reserves (which include condensate and natural gas liquids) shall be stated in barrels, and gas reserves in cubic feet.

(6) *Disclosure of future net revenues from estimated production of proved oil and gas reserves.* In conjunction with the disclosure of changes in net quantities of estimated proved reserves of crude oil (including condensate and natural gas liquids) and natural gas as required by paragraph (k)(5) of this section, the following information shall be disclosed in financial statements for each geographic classification for which quantities of oil and gas are disclosed:

(i) For each of the following categories: (A) Proved oil and gas reserves, (B) proved developed oil and gas reserves, (C) proved oil and gas applicable to long-term supply or similar agreements with foreign governments in which the entity acts as producer, and (D) the entity's share of proved oil and gas reserves of investees accounted for by the equity method, an amount (the "Estimated Future Net Revenues"), in the aggregate, computed by applying current prices of oil and gas (with consideration of price changes only to the extent provided by contractual arrangements) to estimated future production of proved oil and gas reserves as of the date of the latest balance sheet presented, less estimated future expenditures (based on current costs) to be incurred in developing and producing the proved reserves, and assuming continuation of existing economic conditions. Present separately the amounts applicable to each of the first three succeeding fiscal years, and the remainder in a single amount. Indicate the basis used to compute current prices and current costs.

(ii) The present value of the Estimated Future Net Revenues (the "Present Value of Estimated Future Net Revenues"), as of the end of each fiscal year for which an income statement is required, computed using the Estimated Future Net Revenues and a discount factor of ten percent. In addition, disclose the portions of the Present Value of Estimated Future Net Revenues, at such balance sheet dates, attributable to each of the following components of proved oil and gas reserves: (A) Proved reserves added in years prior to the current year; and (B) proved reserves added during the current year, with amounts attributable to application of improved recovery techniques or purchases of reserves-in-place disclosed separately, if significant (which in this case shall generally be considered to be amounts greater than ten percent of total proved reserves added during the current year).

(iii) The Present Value of Estimated Future Net Revenues, computed as set forth at paragraph (k)(6)(i) of this section, attributable to each of the following, stated separately: (A) Proved developed reserves of oil and gas, (B) proved reserves of oil and gas applicable to long-term supply agreements with foreign governments in which the entity acts as producer, and (C) the entity's share of proved reserves of oil and gas of investees accounted for by the equity method.

(iv) Any additional information of which management is aware and which it believes is necessary to prevent the above information from being misleading.

(v) Information relating to the disclosure of future net revenues from estimated production of proved oil and gas reserves pursuant or supplemental to the requirements of this section shall be deemed not to be an untrue statement of a material fact; a statement false or misleading with respect to any material fact; an omission to state a material fact necessary to make a statement not misleading; or the employment of a manipulative, deceptive, or fraudulent device, contrivance, scheme, transaction, act, practice, course of business or an artifice to defraud; as those terms are used in the Securities Act of 1933, the Securities Exchange Act of 1934, or the Public Utility Holding Company Act of 1935, or rules and regulations thereunder, unless such information: (A) was prepared without a reasonable basis; or (B) was disclosed other than in good faith.

* * * * *

These amendments are adopted pursuant to authority in sections 6, 7, 8, 10 and 19(a) (15 U.S.C. 77f, 77g, 77h, 77j, 77s) of the Securities Act of 1933; sections 12, 13, 15(d) and 23(a) (15 U.S.C. 781, 78m, 78o(d), 78w), of the Securities Exchange Act of 1934; sections 5(b), 14, and 20(a) (15 U.S.C. 79c, 79n, 79t) of the Public Utility Holding Company Act of 1935; sections 8, 30, 31(c) and 38(a) (15 U.S.C. 80a-8, 80a-29, 80a-30(c), 80a-37(a)) of the Investment Company Act of 1940; and section 503 (42 U.S.C. 6383) of the Energy Policy and Conservation Act of 1975.

Pursuant to Section 23(a)(2) of the Securities Exchange Act, the Commission has considered the impact of these amendments on competition and is not aware of any burden that they would impose on competition.

By the Commission.

GEORGE A. FITZSIMMONS,
Secretary.

DECEMBER 19, 1978.

[FR Doc. 78-35867 Filed 12-26-78; 8:45 am]

some Dinn 83

SEC Full Cost Rules
Accounting Series Release No. 258

[8010-01-M]

[Release Nos. 33-6007; 34-15417; 35-20837; IC-10531; AS-258]

PART 210 — FORM AND CONTENT OF FINANCIAL STATEMENTS, SECURITIES ACT OF 1933, SECURITIES EXCHANGE ACT OF 1934, PUBLIC UTILITY HOLDING COMPANY ACT OF 1935, INVESTMENT COMPANY ACT OF 1940, AND ENERGY POLICY AND CONSERVATION ACT OF 1975

Oil and Gas Producers — Full Cost Accounting Practices

AGENCY: Securities and Exchange Commission.

ACTION: Final rules.

SUMMARY: The Commission is adopting rules that establish uniform requirements for financial accounting and reporting practices of oil and gas producers following the full cost method of accounting. The Commission has previously published, in Accounting Series Release No. 253, accounting measurement standards for the successful efforts method of accounting. Proposed rules for full cost accounting were published concurrently with that release. The rules being adopted contain modifications based upon evaluation of written comments.

EFFECTIVE DATE: The rules published in this release shall be effective initially for fiscal years ending after December 25, 1979.

FOR FURTHER INFORMATION CONTACT: James L. Russell or Rita J. Gunter, Office of the Chief Accountant, Securities and Exchange Commission, 500 North Capitol Street, Washington, D.C. 205-19 (202-755-0222).

[1]Securities Act Release No. 33-5966 (43 FR 40688).

SUPPLEMENTARY INFORMATION: In Accounting Series Release No. 253 ("ASR No. 253"),[1] issued on August 31, 1978, the Commission announced its conclusions concerning financial accounting and reporting standards for oil and gas producing companies. Among these was the determination that the development of new measurement standards was necessary in order to achieve meaningful reporting of earnings and financial position for companies in that industry. These new standards should provide for recognition in financial statements of proved oil and gas reserves as assets and changes in proved reserves in earnings. In consideration of the need to establish standards for reserve valuations and to achieve an acceptable degree of reliability, a series of stops covering several years was outlined seeking the development of "reserve recognition accounting." During this interim period, the Commission concluded that companies should be permitted to continue to follow either of the two traditional historical cost accounting methods. Rules relating to the application of successful efforts accounting (which conformed to the standards of Statement No. 19 of the Financial Accounting Standards Board) were adopted in ASR No. 253, together with definitions, tax accounting standards, and disclosure requirements applicable to all oil and gas producers.[2] Proposed rules that would add §210.3-18(i) to Regulation S-X to govern the application of full cost accounting for companies following that method were published concurrently in Release No. 33-5968 (43 FR 40724). August 31, 1978.

In that release, the Commission stated that its intention was to develop a practical approach for applying the full cost method of accounting that would be logically consistent and would not differ substantially from existing practice. Comments on the proposal were received from 48 persons, including oil and gas producers, accounting firms, and financial analysts. These comments have been considered by the Commission and various changes are reflected in the rules presently being adopted.

[2]Various technical amendments to these definitions and rules have been adopted by the Commission in Accounting Series Release No. 257, dated December 19, 1978 and published concurrently with this release.

Comments and Changes

COSTS TO BE AMORTIZED

Most letters commenting on the rules proposed in Release No. 33-5968 addressed the issue of whether all capitalized costs should be subject to immediate amortization. It was generally acknowledged that companies with ongoing exploration programs typically have a recurring level of preproduction costs (in relation to the total investment in the cost center), consisting of the cost of undeveloped leases and the related carrying and evaluation costs. This normal inventory of preproduction costs should be included in the cost center for amortization purposes as the costs are incurred, and any additional reserves should be taken into consideration upon discovery on a prospective basis. This method of accounting should result in a reasonable and consistent allocation of costs over a period of time.

However, there was general disagreement with the proposed requirement that the capitalized costs of all undeveloped leases and uncompleted wells and facilities be amortized. It was pointed out that situations may arise in which major expenditures are required prior to production, such as unusually large investments in off-shore leases, major development projects, or secondary and tertiary recovery projects. While previous methods of determining the exclusion of costs from the amortization base have varied among companies, the principle of excluding major costs of this nature was felt by these commentators to be generally accepted. Failure to exclude such costs, it was argued, would result in an unrealistic overstatement of the amortization provision in the early lives of the properties. Exclusion from the amortization base was further supported by the assertion that costs incurred should be considered to be at least equal to the value of the asset at the time of the initial investment. Although certain projects will ultimately be abandoned, it was argued that impairments should be recognized when they occur, and should not be anticipated by premature amortization of the costs. Costs of projects that ultimately prove to be productive should become subject to amortization at the time that this evaluation is determined and the related proved reserves are included in the amortization base.

The Commission considers these arguments to have merit and has modified the rules to provide that the costs of unusually significant acquisitions of unproved properties and major development projects may be excluded from the capitalized costs to be amortized. Specifically, amortization of the costs of unproved properties that are unusually significant (such as the costs of acquiring and evaluating major offshore leases) may be deferred until a determination is made as to whether proved reserves are attributable to the properties, subject to periodic assessment of whether impairment has occurred. Also, the costs of major development projects may be excluded from amortization in cases where unusually significant costs must be incurred prior to a determination of the quantities of proved reserves attributable to the properties under development. Examples of such projects include the installation of an offshore drilling platform from which development wells are to be drilled and improved recovery programs undertaken in expectation of significant additions to proved reserves.

Because of the various judgments involved in assessing significance, the rules do not contain a specific test to determine whether an investment is "unusually significant." However, the Commission has instructed its staff to investigate existing practices and determine whether it is necessary to publish a guideline in this area, such as the percentage of total unamortized capitalized costs that must be exceeded if a major investment project is to be excluded from amortization. It should also be noted that the rules require separate disclosure on the balance sheet of the amount of preproduction costs that are not being amortized and a description of the current status of the projects involved, including an estimate of when the costs will be included in the amortization computation and, to the extent possible, the potential future impact on the amortization rate.

Several commentators questioned the appropriateness of including estimated future development costs in the costs to be amortized, since additions to proved reserves could be expected from major development expenditures that would not be reflected in the amortization computation. The Commission believes that future development costs should be included; the exclusion of the costs of major development projects from the amortization base until reserves can be evaluated should eliminate situations where an inappropriate allocation of costs to current production would result.

BASIS OF AMORTIZATION

Release No. 33-5968 proposed that unit-of-production amortization of capitalized costs be computed either on the basis of physical units of oil and gas (converted to a common unit of measure according to their relative energy content) or monetary units of net revenue. Several commentators recommended that only the use of physical units should be permitted; this would be consistent with the requirements for amortization of costs in cases of joint production of oil and gas under the Commission's successful efforts rules. Other commentators expressed agreement with the proposal to permit use of revenue units, since this computation utilizes the same basic information as amortization computations involving physical units, while automatically compensating for differences in the relative sales prices of the reserves. These commentators also pointed out that a method based on energy content ignores the fact that prices of oil and gas in today's regulated market environment do not necessarily correspond to their relative energy content.

However, the commentators favoring revenue units generally indicated that the computation should be based on gross revenues rather than net revenues. The use of net revenues, it was argued, requires estimates of future costs and introduces distortions when unusually high or low operating expenses are incurred in individual years. On the other hand, the relevant information for gross revenues is readily attainable, since current production is already recorded in dollars as revenues, and reserve reports frequently indicate reserves in terms of dollars based on current prices.

The Commission believes that there should be consistency in the methods used by companies following the full cost method of accounting. Further, the Commission believes that the use of physical units in terms of energy content (BTU's) as the amortization basis has the greatest conceptual merit in a market environment in which the relative prices of oil and gas approximate their relative energy content. However, the fact that substantial production is currently subject to pricing regulations may result in significant differences in amortization under the BTU method as compared to amortization based on sales values. Accordingly, the final rules prescribe the use of the physical units method unless the use of such method, because of existing pricing regulations, would cause an amortization provision that would be inconsistent with the current prices being received. In those instances, the gross revenue method may be used as the basis for computing amortization rates.

LIMITATION ON CAPITALIZED COSTS

A majority of commentators expressed reservations concerning the proposal that the cost center ceiling be based on the present value of future net revenues from estimated production of proved oil and gas reserves. It was stated that future net revenues should not be discounted for purposes of the ceiling limitation, that other classifications of reserves in addition to proved reserves should be considered, or that anticipated price increases should be included. The primary concern expressed by these commentators was that the proposed ceiling computation, in certain cases, would result in write-offs that would reduce net capitalized costs below the actual realizable value of the properties.

The Commission continues to believe that the aggregate of (1) the present value of proved reserves and (2) the lower of cost or fair market value of unproved properties is an appropriate basis for determining the limitation on capitalized costs. However, the rules have been revised to specify that the costs of significant investments in unproved properties and development projects are to be considered separately for purposes of the ceiling computation.

If application of the rules, as a result of an unusual event or transaction such as a major purchase of proved properties, would require a write down when the fair value of the properties in a cost center clearly exceeds the unamortized costs, the registrant may request an exemption from the general rule. In such cases, the registrant should be prepared to demonstrate that the additional value clearly exists beyond reasonable doubt.

Finally, the rules specifiy that the cost center ceiling is to be computed giving consideration to income tax effects. Comments were received indicating that such income tax effects would be difficult to estimate and, in most cases, would not have a significant effect on the ceiling computation. However, the Commission believes that unusual tax relationships may exist in certain instances, as a result of the expiration of operating loss carryforwards, changes in tax rates, etc. In these circumstances, it will be necessary to consider tax effects in computing the ceiling limitation.

MINERAL PROPERTY CONVEYANCES

The rules dealing with mineral property conveyances specify that a sale of oil and gas reserves shall be accounted for as an adjustment of capitalized costs, unless the adjustment causes a significant alteration of the relationship between remaining capitalized costs and proved reserves attributable to the cost center. Guidance has been added that a significant alteration would not normally be expected to result from a sale involving less than 25 percent of the total reserve quantities of the cost center.

Paragraph (i)(6)(i) of the proposed rules would have required, in cases where the recognition of gain or loss on a sale of oil and gas properties is appropriate, that capitalized costs be allocated between reserves sold and reserves retained on the same basis used to compute amortization. Several commentators pointed out that the value of proved reserves sold and retained could differ significantly as the result of price regulations or quality differences. The sales price might also reflect values other than proved reserves, such as probable reserves or unproved leases, which should also be considered in the cost allocation. The rules as adopted provide that costs should be allocated between cost center assets sold and retained in the manner described in the proposed rules except where estimated relative fair values should be used in order to reflect other substantial economic differences between the properties sold and those retained.

The accompanying rules also clarify the Commission's intent that income from sales of unproved properties or from drilling arrangements should be recognized only in unusual situations where the cash consideration received exceeds the total cost of the properties plus any exploration and development costs to be subsequently incurred. Otherwise, income from sales of unproved properties or from drilling arrangements with respect to properties should be credited to the cost center, and any related costs should be capitalized as part of that cost center. In this way, drilling arrangement income is recognized as a reduction of overall costs of oil and gas reserves. The accounting for drilling arrangements should reflect the substance of the transactions, regardless of the form of the underlying agreements, including consideration of guarantees of performance or indebtedness.

DISCLOSURES

The Commission's proposed rules called for the disclosure by geographic area of (1) the approximate amount of capitalized costs that would have been charged to expense had the successful efforts method of accounting been followed, and (2) the approximate amount of costs incurred that would have been charged to expense under the successful efforts method. Many commentators objected to this proposed requirement as an unreasonable and unnecessary burden. It was also seen by these persons as inconsistent with the basic conclusions of ASR No. 253 with respect to the severe limitations of both historical cost methods in permitting investors and government policy-makers to "gain an understanding of the operations of individual companies or to compare the operations of different companies." A small minority, however, expressed the view that the "as if" disclosures were essential for comparing the financial position and operating results of companies using alternative accounting methods during the period of implementation of reserve recognition accounting. Those in favor of the supplemental disclosures would, in addition, expand the requirement to include disclosure of pro forma net income.

The Commission has decided to defer temporarily any final action on this proposed disclosure requirement until it considers various questions that have been raised with respect to accounting changes and preferability of accounting methods in this industry. Therefore, the accompanying rules do not contain a requirement for supplemental disclosures based on successful efforts accounting, and this disclosure rule proposal remains outstanding.

CONSOLIDATED FINANCIAL STATEMENTS

Release No. 33-5968 stated that a company following the full cost method would be required to apply that method to all of its operations, including those of its subsidiaries and of investees accounted for by the equity method. Commentators pointed out that the requirement regarding equity investees could present substantial difficulties. If an equity investee were following the successful efforts method, the information necessary to convert a company's investment would not be available on a timely basis in many situations. The Commission has reconsidered this proposal and has eliminated the requirement that the investor's accounting method be applied to its equity investees. However, the final rules specify that a registrant must apply its accounting method to the operations of its subsidiaries.

A number of commentators also requested clarification of other requirements as they apply to companies within a consolidated group. Paragraph (i)(3)(v) requires that amortization rates be determined on a consolidated basis even though this may result in a consolidated amortization provision that is not equal to the sum of the expenses for the individual members of the consolidated group. This same concept applies to the determination of the limitation on capitalized costs within cost centers.

RETROACTIVE RESTATEMENT

Several persons suggested that general guidance be provided for the retroactive application of the full cost rules, because of the number of estimates and judgments necessary in making such restatements. Accordingly, a discussion of the initial application of the rules is included in this release. Specifically, any provision of the rules that would not have a significant effect on prior years' financial statements need not be retroactively applied. Estimates of quantities of oil and gas reserves made in prior years for purposes of amortization computations shall not currently be revised in retrospect. Furthermore, if unamortized costs capitalized within a cost center do not exceed the ceiling limitation when the rules are initially applied, it is not necessary to determine whether there may have been an excess in prior years.

OTHER MATTERS

Several comment letters contained suggestions that cost centers of more than one country may be appropriate in certain cases (e.g., U.S. and Canada, or contiguous offshore areas such as the North Sea). However, oil and gas reserves in different countries are not always interchangeable, either in kind or in monetary value, because of varying energy and tax laws, market conditions, property rights, and national economic policies. Therefore, the requirement for establishing cost centers on a country-by-country basis (including related offshore areas) has been retained.

In order to conform to comparable rules for the successful efforts method, paragraph (i)(3)(iv) has been added to acknowledge that, in some cases, depreciation of natural gas cycling and processing plants may appropriately be computed by a method other than unit-of-production.

Finally, a number of commentators expressed concern with the applicability of the proposed rules to properties regulated for rate-making purposes on an individual-company-cost-of-service basis. Among the technical amendments to its rules on financial accounting and reporting by oil and gas producers adopted by the Commission in Accounting Series Release No. 257 is a provision that appropriate recognition may be given to differences in capitalized costs and in the basis for amortization arising because of the effect of the rate-making process. This applies to both the successful efforts and the full cost rules. However, the Commission believes that the disclosure requirements of §210.3-18(k) are relevant to rate-regulated companies, and therefore has not exempted them from these disclosure requirements.

Effective Date and Initial Application

The provisions of §210.3-18 are effective for fiscal years ending after December 25, 1978, that are contained in filings that include fiscal years ending after December 25, 1979, although earlier application is encouraged. Accounting changes adopted to conform to these rules are to be made retroactively by restating the financial statements of prior periods. Financial statements for the fiscal year in which these rules are first applied should disclose the nature of the accounting changes and their effect on income before extraordinary items, net income, and related per share amounts for each period restated.

Retroactive application of the provisions of these rules requires the use of estimates and approximations. A provision that would not have a significant effect on prior years' financial statements need not be retroactively applied. Further, retroactive application of some provisions of these rules may require the use of estimates of a type not previously made; information that may have become available some time after the year being restated may be taken into account in making those estimates, except that estimates of quantities of oil and gas reserves that had been made in prior years shall not currently be revised in retrospect.

For reporting entities following the full cost method of accounting, retroactive application of §210.3-18(i)(4), "Limitations on capitalized costs," shall be applied as follows:

(a) If unamortized costs capitalized within a cost center do not exceed the cost center ceiling as of the beginning of the fiscal period in which the rules are initially adopted, then no provisions shall be made for past periods when application of the rules based on information known during those periods might have resulted in unamortized capitalized costs being in excess of the cost center ceiling.

(b) If unamortized costs capitalized within a cost center exceed the cost center ceiling as of the beginning of the fiscal year in which the rules are initially adopted, then this excess shall be recognized retroactively through a charge to expense in the periods in which the excess initially arose.

COMMISSION ACTION

The Commission hereby amends 17 CFR Part 210 by the addition of new paragraph (i) to §210.3-18, as set forth below:

§210.3-18 Financial accounting and reporting for oil and gas producing activities pursuant to the federal securities laws and the Energy Policy and Conservation Act of 1975.

* * * * *

(i) *Application of the full cost method of accounting.* A reporting entity that follows the full cost method shall apply that method to all of its operations and to the operations of its subsidiaries, as follows:

(1) *Determination of cost centers.* Cost centers shall be established on a country-by-country basis.

(2) *Costs to be capitalized.* All costs associated with property acquisition, exploration, and development activities (as defined in paragraph (a) of this section) shall be capitalized within the appropriate cost center. Any internal costs that are capitalized shall

be limited to those costs that can be directly identified with acquisition, exploration, and development activities undertaken by the reporting entity for its own account, and shall not include any costs related to production, general corporate overhead, or similar activities.

(3) *Amortization of capitalized costs.* Capitalized costs within a cost center shall be amortized on the unit-of-production basis using proved oil and gas reserves, as follows:

(i) Costs to be amortized shall include (A) all capitalized costs, less accumulated amortization, other than the cost of properties described in paragraph (ii) below; (B) the estimated future expenditures (based on current costs) to be incurred in developing proved reserves; and (C) estimated dismantlement and abandonment costs, net of estimated salvage values.

(ii) The cost of unusually significant investments in unproved properties and major development projects may be excluded from capitalized costs to be amortized, subject to the following:

(A) Costs of acquiring and evaluating unproved properties may be excluded only if the costs incurred are unusually significant in relation to the aggregate costs to be amortized (e.g., the costs of acquiring major offshore leases). All costs of acquiring such properties and related exploration costs shall be excluded from the amortization computation until it is determined whether or not proved reserves are attributable to the properties. Until such a determination is made, the properties shall be assessed individually to ascertain whether impairment has occurred (see paragraph (c) of this section). If the results of the assessment indicate impairment, the amount of the impairment shall be added to the costs to be amortized.

(B) Costs of major development projects may be excluded from amortization only if unusually significant development costs must be incurred prior to ascertaining the quantities of proved reserves attributable to the properties under development (e.g., the installation of an offshore drilling platform from which development wells are to be drilled, the installation of improved recovery programs, and similar major projects undertaken in the expectation of significant additions to proved reserves). In such cases, a portion of the development costs identified with such a project may be excluded from the costs to be amortized until the proved reserves added as a result of the project are ascertainable or until it is determined that impairment has occurred.

(iii) Amortization shall be computed on the basis of physical units, with oil and gas converted to a common unit of measure on the basis of their approximate relative energy content, unless economic circumstances (related to the effects of regulated prices) indicate that use of units of revenue is a more appropriate basis of computing amortization. In the latter case, amortization shall be computed on the basis of current gross revenues (excluding royalty payments and net profits disbursements) from production in relation to future gross revenues, based on current prices (including consideration of changes in existing prices provided only by contractual arrangements), from estimated production of proved oil and gas reserves. The effect of a significant price increase during the year on estimated future gross revenues shall be reflected in the amortization provision only for the period after the price increase occurs.

(iv) In some cases it may be more appropriate to depreciate natural gas cycling and processing plants by a method other than the unit-of-production method.

(v) Amortization computations shall be made on a consolidated basis, including investees accounted for on a proportionate consolidation basis. Investees accounted for on the equity method shall be treated separately.

(4) *Limitation on capitalized costs:* (i) For each cost center, capitalized costs, less accumulated amortization and related deferred income taxes, shall not exceed an amount (the cost center ceiling) equal to the sum of: (A) The present value of future net revenues from estimated production of proved oil and gas reserves as defined in paragraph (k)(6) of this section; plus (B) the cost of properties not being amortized pursuant to paragraph (i)(3)(ii) of this section; plus (C) the lower of cost or estimated fair value of unproved properties included in the costs being amortized; less (D) income tax effects related to differences between the book and tax basis of the properties involved.

(ii) If unamortized costs capitalized within a cost center, less related deferred income taxes, exceed the cost center ceiling, the excess shall be charged to expense and separately disclosed during the period in which the excess occurs. Amounts thus required to be written off shall not be reinstated for any subsequent increase in the cost center ceiling.

(5) *Production costs.* All costs relating to production activities, including workover costs incurred solely to maintain or in-

crease levels of production from an existing completion interval, shall be charged to expense as incurred.

(6) *Mineral property conveyances and related transactions. The provisions of paragraph (h) of this section, "Mineral property conveyances and related transactions if the successful efforts method of accounting is followed," shall apply also to those reporting entities following the full cost method except as follows:*

(i) Sales and abandonments of oil and gas properties. Sales of oil and gas reserves in place and abandonments of properties shall be accounted for as adjustments of capitalized costs, with no gain or loss recognized, unless such adjustments would significantly alter the relationship between capitalized costs and proved reserves of oil and gas attributable to a cost center. A significant alteration would not ordinarily be expected to occur for sales involving less than 25 percent of the reserve quantities of a given cost center. If gain or loss is recognized for such a sale, total capitalized costs within the cost center shall be allocated between the reserves sold and reserves retained on the same basis used to compute amortization, unless there are substantial economic differences between the properties sold and those retained, in which case capitalized costs shall be allocated on the basis of the relative fair values of the properties.

(ii) *Purchases of reserves.* Purchases of oil and gas reserves in place ordinarily shall be accounted for as additional capitalized costs within the applicable cost center; however, significant purchases of production payments or properties with lives substantially shorter than the composite productive life of the cost center shall be accounted for separately.

(iii) *Drilling arrangements.* Consistent with the provisions of paragraph (h) of this section, no income shall be recognized from sales of unproved properties or participation in various forms of drilling arrangements involving oil and gas producing activities (e.g., carried interest, turnkey wells, management fees, etc.) if the sales of unproved properties or drilling arrangements related thereto, in substance, will provide for the receipt or retention of an economic interest in any form in the properties. Circumstances in which income may be recognized are limited to situations where (A) the cash consideration received from the sale of unproved properties or drilling arrangements involving oil and gas producing activities exceeds the total cost of the properties plus any exploration and development costs to be subsequently incurred, or (B) the cash compensation represents reimbursement for amounts currently charged to expense. In the case of partnership or joint venture operations undertaken by the entity that involve more than one property, the determination of whether the cash compensation exceeds the cost of the properties shall be made based on the entity's participation in the total operations of the partnership or joint venture, rather than on a property-by-property basis. If the cash consideration received represents reimbursement for organization, offering, general and administrative expenses, etc., such compensation may be recognized as income only to the extent that costs have been currently incurred and charged to expense.

(7) *Disclosures.* Reporting entities that follow the full cost method of accounting shall disclose all of the information required by paragraph (k) of this section, with each cost center considered as a separate geographic area, except that reasonable groupings may be made of cost centers that are not significant in the aggregate. In addition:

(i) For each cost center for each year that an income statement is required, disclose the total amount of amortization expense (per equivalent physical unit of production if amortization is computed on the basis of physical units or per dollar of gross revenue from production if amortization is computed on the basis of gross revenue).

(ii) State separately on the face of the balance sheet the aggregate of the capitalized costs of unproved properties and major development projects that are excluded, in accordance with paragraph (i)(3) of this section, from the capitalized costs being amortized. Provide a description in the notes to the financial statements of the current status of the significant properties or projects involved, including the anticipated timing of the inclusion of the costs in the amortization computation and, to the extent possible, the potential future impact on the amortization rate. In addition, indicate the nature of the costs by category and the approximate date on which costs were first incurred with respect to each such property or project.

* * * * *

These amendments are adopted pursuant to authority in sections 6, 7, 8, 10 and 19(a) (15 U.S.C. 77f, 77g, 77h, 77j, 77s) of the

Securities Act of 1933; sections 12, 13, 15(d), and 23(a) (15 U.S.C. 78l, 78m, 78o(d), 78w) of the Securities Exchange Act of 1934; section 5(b), 14, and 20(a) (15 U.S.C. 79e, 79n, 79t) of the Public Utility Holding Company Act of 1935; sections 8, 30, 31(c) and 38(a) (15 U.S.C. 80a-8, 80a-29, 80a-30(c), 80a-37(a)) of the Investment Company Act of 1940; and section 503 (42 U.S.C. 6383) of the Energy Policy and Conservation Act of 1975.

Pursuant to section 23(a)(2) of the Securities Exchange Act, the Commission has considered the impact that these rules might have on competition and has concluded that to the extent the rules impose burdens on competition, such burdens are necessary and appropriate in furtherance of the purposes of the securities laws.

By the Commission.

GEORGE A. FITZSIMMONS,
Secretary.

DECEMBER 19, 1978.

[FR Doc. 78-35868 Filed 12-26-78; 8:45 am]

Appendix B

Examples of Oil and Gas Producing Company Financial Statements.

Part I of this appendix is excerpts from the 1981 annual report to stockholders of Basic Earth Science Systems, Inc., a successful efforts company.

Part II of this appendix is excerpts from the 1981 annual report to stockholders of Amarex, Inc., a full costing company.

Part I — Basic Earth Science Systems, Inc.

Selected Financial Data
Basic Earth Science Systems, Inc.

Year ended March 31,	**1981**	1980	1979	1978	1977
Oil and gas sales	**$ 3,279,000**	$ 2,650,000	$1,269,000	$ 1,552,000	$ 1,560,000
Oil and gas operations expense	**813,000**	382,000	792,000	523,000	452,000
Exploration expense	**1,334,000**	177,000	27,000	236,000	7,000
Net income (loss) from continuing operations	**(943,000)**	382,000	(741,000)	(77,000)	181,000
Earnings (loss) per share from continuing operations (2)	**(.09)**	.04	(.08)	(.01)	.02
Total Assets	**18,815,000**	8,406,000	8,279,000	11,921,000	11,111,000
Long term debt	**496,000**	2,920,000	4,373,000	4,172,000	2,842,000
Total oil and gas reserves on basis of RRA (1)	**62,004,000**	106,030,000	3,322,000	*	*
Earnings (loss) per share from oil and gas operations on basis of RRA (1) (2)	**(2.93)**	6.77	*	*	*

No cash dividends have ever been paid by Basic.

*Information on basis of RRA was not prepared and is not available for the years ended March 31, 1979 and earlier.

(1) See footnote 14 to the consolidated financial statements for an explanation of Reserve Recognition Accounting (RRA).

(2) The EPS figures have been retroactively adjusted to reflect the stock split declared in May, 1981.

Management's Discussion of Financial Condition and Results of Operations

Liquidity and Capital Resources
At March 31, 1981, Basic's current ratio was 2.5 to 1. This is the most favorable current ratio in Basic's history. This is the result of Basic's secondary public offering closed in January, 1981. Basic received $14,577,000 net proceeds from this offering. Those proceeds were used to pay off the majority of Basic's long-term debt with the remainder to working capital for developmental and exploratory drilling and other corporate purposes.

Basic expects to spend $11,250,000 on drilling 49 wells during the fiscal year ending March 31, 1982. This will be financed from the remainder of the proceeds of the offering and a $10,000,000 revolving note obtained in May, 1981. This note is payable in equal monthly principal payments over a four year period beginning in April, 1982.

Results of Operations
The year ended March, 1981 was the first full year that Basic concentrated all its efforts in one industry segment, oil and gas exploration and production. Basic's petroleum services segment was discontinued in January, 1980.

In 1981 and 1980, oil and gas sales increased over the prior years. This was due mainly to new oil and gas producing properties and increases in price. In 1981, new properties made up $479,000 of the increase over the prior year while in 1980 new properties accounted for $990,000 of the increase. The price of newly discovered oil in 1980 was significantly greater than in 1979, but in 1981 the price dropped. In 1981, the price of old oil (oil discovered prior to 1978) was decontrolled and thus began to rise to the market price of newly discovered oil. Newly discovered oil was selling for approximately $41.00 per barrel in March, 1980 compared to approximately $36.00 per barrel in March, 1981.

Oil and gas operations expense in 1981 increased substantially over the amount in 1980. This was primarily caused by a full year of operations of the West Cole Unit and West Cole North Unit waterflood projects, since application of secondary recovery techniques greatly increases the operating costs of an oil and gas property. The increase in operating costs of these two waterflood projects accounted for 56% of the increase in oil and gas operations over 1980. New wells put on production in 1981 accounted for 32% of this increase.

There was a large increase in exploration expense in 1981 as a result of Basic's drilling a greater number of exploratory wells than in the past. Exploration expense is the cost of dry exploratory wells drilled during the period and determined to be dry. Although the risk of drilling an exploratory well is higher than a development well, the potential reward is also greater. Basic intends to spend over $11 million drilling wells in 1982, partially financed by the proceeds of the secondary public offering completed in January, 1981. This is the largest drilling budget in the Company's history.

Oil and Gas Operations

	Revenue					Expense
	Oil		Gas			
	Barrels	Amount	Thousand Cubic Feet	Amount	Total	
Year ended 3/31/81	73,000	$2,489,000	468,000	$790,000	$3,279,000	$813,000
Year ended 3/31/80	76,000	$2,080,000	376,000	$570,000	$2,650,000	$382,000
Year ended 3/31/79	58,000	$ 748,000	433,000	$521,000	$1,269,000	$792,000

Management's Discussion of Financial Condition and Results of Operations

General and administrative expense increased significantly over the last three years. The largest portion of this increase was the general and administrative costs which were allocated to the discontinued petroleum services segment in prior periods. These costs were partially absorbed by the oil and gas exploration segment in 1980 and were fully absorbed by the oil and gas exploration segment in 1981. The increase in drilling activity mentioned above has required an expanded support effort which also increased general and administrative expense.

Beginning March 1, 1980, the federal government imposed the "Windfall Profit" excise tax on domestic sales of crude oil. The tax does not apply to income earned from the sale of natural gas. On each barrel of taxable crude oil, the tax equals the "windfall profit" multiplied by the applicable tax rate. The "windfall profit" is the difference between the selling price of a barrel of oil and the adjusted base price, which will vary depending upon the tax category in which the oil falls, less an adjustment for state severance taxes. Independent producers are provided with reduced tax rates on their production which does not exceed 1,000 barrels a day averaged over a calendar quarter. Because Basic is an independent producer with less than 1,000 barrels a day production, most of its production is taxed at a 30% rate. The year ended March 31, 1981 was the first full year that Basic was subject to this tax. Basic paid an average of $2.22 per barrel in "Windfall Profit" tax.

Market for Common Stock and Dividend Information

In May, 1981, Basic declared a split of Basic's common stock. On May 26, 1981 an extra share was distributed to all shareholders of record on May 12, 1981. Immediately after the split there were 12,099,266 shares outstanding. The schedule on page one has been retroactively adjusted to reflect this stock split.

In January, 1981, Basic completed a secondary public offering of common stock. Basic sold 1,100,000 units at $15.00 per unit. Each unit consisted of one share of common stock and one warrant to purchase one share of common stock. As a result of the stock split in May, 1981, each warrant can be used to purchase two shares of common stock.

In June, 1981, Basic extended the life of the warrants so that each warrant entitles the holder to purchase two shares of common stock at $15.75 ($7.875 per share) until June 17, 1982 and $18.00 ($9.00 per share) from June 18, 1982 to June 17, 1983.

As part of the above sale of stock, Basic sold to the underwriter, warrants to purchase 400,000 shares of common stock. These warrants can be exercised at a price of $9.00 per share from December 18, 1981 to December 17, 1985.

The number of holders of record of Basic's common stock as of April 23, 1981 was 1,602.

Basic has paid no dividends on its common stock since its inception other than two stock dividends totaling 900,770 shares issued in 1977. It is the philosophy of management that the return on shareholders' investments will be maximized by reinvesting all earnings in the exploration for and discovery of additional oil and gas reserves. Accordingly, Basic intends to follow a policy of retaining all earnings to provide funds for expansion of its business and does not anticipate paying any cash dividends in the near future.

Consolidated Statement of Operations

Basic Earth Science Systems, Inc.

17

Year ended March 31,	**1981**	1980	1979
Revenue:			
Oil and gas sales	**$3,279,000**	$ 2,650,000	$ 1,269,000
Interest income	**211,000**	—	—
Gain on sale of assets	**98,000**	—	—
Realized gain on sale of marketable securities	**15,000**	—	—
	3,603,000	2,650,000	1,269,000
Expense:			
Oil and gas operations	**813,000**	382,000	792,000
Exploration	**1,334,000**	177,000	27,000
Depreciation, depletion and amortization	**668,000**	409,000	699,000
Interest	**546,000**	306,000	215,000
General and administrative	**782,000**	306,000	178,000
Other taxes and licenses	**241,000**	200,000	99,000
Windfall Profit tax	**162,000**	32,000	—
	4,546,000	1,812,000	2,010,000
Income (loss) from continuing operations before income tax	**(943,000)**	838,000	(741,000)
Income tax	—	(456,000)	—
Income (loss) from continuing operations	**(943,000)**	382,000	(741,000)
Discontinued operations:			
Loss from operations of discontinued segments, less income tax recovery of $456,000 in 1980	—	(812,000)	(2,260,000)
Loss on disposal of segments	—	(781,000)	(98,000)
	—	(1,593,000)	(2,358,000)
Net loss	**$ (943,000)**	$(1,211,000)	$(3,099,000)
Weighted average number of shares outstanding	**10,432,600**	9,854,140	9,729,216
Earnings (loss) per share:			
Earnings (loss) per share from continuing operations	**$(.09)**	$ .04	$(.08)
Loss per share from discontinued operations	—	(.16)	(.24)
Loss per share	**$(.09)**	$(.12)	$(.32)

The accompanying notes are an integral part of the consolidated financial statements.

Consolidated Balance Sheet

Basic Earth Science Systems, Inc.

Assets		
March 31,	**1981**	1980
Current assets:		
Cash	**$ 92,000**	$ 271,000
Invested cash	**5,000,000**	—
Marketable securities, at aggregate cost (market value — $1,661,000)	**828,000**	—
Accounts and notes receivable (net of allowance for doubtful accounts of $398,000 in 1981 and $123,000 in 1980)	**1,368,000**	568,000
Other	**70,000**	—
Assets of discontinued segments at estimated net realizable value	**175,000**	2,826,000
Total current assets	**7,533,000**	3,665,000
Property and equipment:		
Oil and gas properties (successful efforts)	**13,478,000**	6,762,000
Support equipment and other	**946,000**	347,000
	14,424,000	7,109,000
Less accumulated depreciation, depletion and amortization	**3,142,000**	2,464,000
	11,282,000	4,645,000
Other assets	—	96,000
	$18,815,000	$8,406,000

Liabilities and Shareholders' Equity		
Current liabilities:		
Accounts payable and accrued expenses	**$ 2,358,000**	$1,869,000
Current portion of long-term debt	**632,000**	1,993,000
Total current liabilities	**2,990,000**	3,862,000
Long-term debt	**496,000**	2,920,000
Contingencies and commitments	—	—
Shareholders' Equity:		
Preferred stock, $.10 par value Authorized — 3,000,000 shares Issued — 0 shares in 1981	—	—
Common stock, $.10 par value Authorized — 32,000,000 shares Issued — 12,235,324 shares in 1981 and 10,010,138 shares in 1980	**1,223,000**	1,000,000
Capital in excess of par value	**20,737,000**	6,290,000
Employee Stock Ownership Plan contribution	—	50,000
Accumulated deficit since April 1, 1974	**(6,431,000)**	(5,488,000)
	15,529,000	1,852,000
Less treasury stock, at cost (136,058 shares in 1981 and 156,058 shares in 1980)	**200,000**	228,000
	15,329,000	1,624,000
	$18,815,000	$8,406,000

The accompanying notes are an integral part of the consolidated financial statements.

Consolidated Statement of Changes in Financial Position

Basic Earth Science Systems, Inc.

Year ended March 31,	**1981**	1980	1979
Financial resources provided by:			
Income (loss) from continuing operations	**$ (943,000)**	$ 382,000	$ (741,000)
Items not affecting working capital:			
Depreciation, depletion and amortization	**668,000**	409,000	699,000
Cost of oil and gas properties sold or abandoned	**1,251,000**	243,000	579,000
Loss (gain) on sales of property and equipment	**(98,000)**	(58,000)	11,000
Employee Stock Ownership Plan contribution	**—**	50,000	20,000
Income tax expense	**—**	456,000	—
Other	**35,000**	(20,000)	18,000
Working capital provided by continuing operations	**913,000**	1,462,000	586,000
Working capital provided by (applied to) discontinued operations	**—**	1,421,000	(366,000)
Working capital provided by operations	**913,000**	2,883,000	220,000
Proceeds of long-term debt	**11,342,000**	4,998,000	4,552,000
Proceeds of sale of property and equipment	**250,000**	157,000	839,000
Reduction of long-term notes receivable	**96,000**	44,000	32,000
Issuance of common stock, common stock warrants and treasury stock	**14,648,000**	—	—
	27,249,000	8,082,000	5,643,000
Financial resources applied to:			
Reduction of long-term debt	**13,766,000**	6,450,000	4,351,000
Purchase of property and equipment	**8,743,000**	2,199,000	1,416,000
Increase in long-term notes receivable	**—**	—	98,000
	22,509,000	8,649,000	5,865,000
Increase (decrease) in working capital	**$ 4,740,000**	$ (567,000)	$ (222,000)
Increase (decrease) in working capital:			
Cash	**$ (179,000)**	$ 127,000	$ (65,000)
Invested cash	**5,000,000**	—	—
Marketable securities	**828,000**	—	—
Accounts and notes receivable	**800,000**	(530,000)	(645,000)
Other assets	**70,000**	(249,000)	17,000
Assets of discontinued segments	**(2,651,000)**	2,826,000	—
Accounts payable	**(489,000)**	(808,000)	(120,000)
Current portion of long-term debt	**1,361,000**	(1,933,000)	591,000
Increase (decrease) in working capital	**$ 4,740,000**	$ (567,000)	$ (222,000)

The accompanying notes are an integral part of the consolidated financial statements.

Consolidated Statement of Changes in Shareholders' Equity
Basic Earth Science Systems, Inc.
Years ended March 31, 1981, 1980 and 1979

	Common stock		Capital in excess of par value	Employee Stock Ownership Plan contribution	Accumulated deficit since April 1, 1974	Treasury stock	
	Shares	Par value				Shares	Amount
Balance, April 1, 1978	9,809,806	$ 980,000	$ 6,194,000	$ 61,000	$(1,178,000)	156,058	$228,000
Issuance of stock in connection with Employee Stock Ownership Plan contribution	75,468	8,000	53,000	(61,000)	—	—	—
Employee Stock Ownership Plan contribution (124,864 shares)	—	—	—	55,000	—	—	—
Net loss	—	—	—	—	(3,099,000)	—	—
Balance, March 31, 1979	9,885,274	988,000	6,247,000	55,000	(4,277,000)	156,058	228,000
Issuance of stock in connection with Employee Stock Ownership Plan contribution	124,864	12,000	43,000	(55,000)	—	—	—
Employee Stock Ownership Plan contribution (25,186 shares)	—	—	—	50,000	—	—	—
Net loss	—	—	—	—	(1,211,000)	—	—
Balance, March 31, 1980	10,010,138	1,000,000	6,290,000	50,000	(5,488,000)	156,058	228,000
Issuance of stock in connection with Employee Stock Ownership Plan contribution	25,186	3,000	47,000	(50,000)	—	—	—
Issuance in public offering (net of offering costs)	2,200,000	220,000	14,357,000	—	—	—	—
Issuance of treasury shares for consulting contract	—	—	43,000	—	—	(20,000)	(28,000)
Net loss	—	—	—	—	(943,000)	—	—
Balance, March 31, 1981	**12,235,324**	**$1,223,000**	**$20,737,000**	**$ —**	**$(6,431,000)**	**136,058**	**$200,000**

The accompanying notes are an integral part of the consolidated financial statements.

Notes to Consolidated Financial Statements

Basic Earth Science Systems, Inc.

Index to Notes to Consolidated Financial Statements

Notes to Consolidated Financial Statements

Basic Earth Science Systems, Inc.

1. Summary of Significant Accounting Policies

This summary of significant accounting policies of Basic Earth Science Systems, Inc. (Basic) is presented to assist in understanding Basic's financial statements. These accounting policies conform to generally accepted accounting principles and have been consistently applied, except for the change in the method of recording interest as described in Note 11.

Principles of Consolidation
The consolidated financial statements include the accounts of Basic and its wholly-owned subsidiaries. All significant inter-company accounts and transactions have been eliminated.

Oil and Gas Producing Activities
Basic follows the successful efforts method of accounting for acquisition, exploration, development and production of its oil and gas properties. Under this method, lease bonuses and commissions paid in connection with the acquisition of leasehold interests are capitalized (acquisition costs), and if undeveloped, assessed annually to determine possible impairment of value. If value is determined to be impaired, the carrying value is reduced by a charge to current operations. Delay rentals paid to retain lease rights are charged to operations. Geological and geophysical costs are expensed as incurred.

The costs of wells-in-progress are deferred until it can be determined if proved reserves have been discovered. All costs of successful wells and dry development wells are capitalized (development costs). Costs associated with dry exploration wells are charged to expense when it is determined that proved reserves have not been discovered.

Basic uses the cost recovery method to record gains (losses) on drilling arrangements which are recorded in operations, net of costs.

Investment Tax Credits
Basic follows the flow-through method of accounting for investment tax credits. Accordingly, the investment tax credits will be reflected in the financial statements as a reduction of income taxes in the year in which they are utilized.

Property Retirements
Repairs and maintenance are charged to expense, and renewals and betterments are capitalized. When property and equipment is retired, abandoned or sold, the related costs and accumulated depreciation and depletion are removed from the accounts and any gain or loss is reflected in operations.

Marketable Securities
Marketable securities are carried at lower of aggregate cost or market value. Realized gains or losses are computed using the specific identification method for the value of securities in the portfolio.

Depreciation, Depletion and Amortization
Depreciation of other than oil and gas property and equipment is computed on the straight-line method over periods ranging from three to thirty years.

Depreciation and depletion on producing oil and gas properties, including equipment, is computed on the unit-of-production method using proved reserves for lease acquisition costs and proved developed reserves for capitalized exploration and development costs.

2. Oil and Gas Properties

The following is a summary of Basic's oil and gas expenditures, all of which are in the United States:

Year ended March 31,	**1981**	1980	1979
Property acquisition costs	**$ 1,713,000**	$ 232,000	$ 24,000
Exploratory costs	**2,427,000**	787,000	71,000
Development costs	**3,884,000**	937,000	132,000
Production costs	**1,180,000**	647,000	545,000
	$ 9,204,000	$2,603,000	$772,000

The aggregate amount of capitalized cost of oil and gas properties was comprised of the following:

March 31,	**1981**	1980
Proved properties	**$10,726,000**	$6,641,000
Unproved properties	**2,752,000**	121,000
	$13,478,000	$6,762,000
Accumulated depreciation, depletion and amortization	**$ 3,056,000**	$2,368,000

2. Oil and Gas Properties (cont'd)

During the years ended March 31, 1981, 1980 and 1979, none of the oil and gas sales were from foreign production or from reserves applicable to long-term supply or similar agreements with foreign governments. In fiscal 1981, 1980 and 1979, respectively, net revenue from oil and gas production (revenue less production, ad valorem and "Windfall Profit" tax and lease operating expense) was $2,099,000, $2,003,000 and $724,000.

3. Long-Term Debt

March 31,	**1981**	1980
1.5% above prime (17.0% prime at March 31, 1981) revolving bank note, payable at maturity with interest payable monthly, secured by certain producing oil and gas properties and, in 1980, by petroleum services equipment, due June 1, 1981	**$ 632,000**	$ 748,000
1.5% above prime bank note (17.0% prime at March 31, 1981), secured by certain producing oil and gas properties and, in 1980, by petroleum services equipment, due July 1984	**10,000**	4,008,000
11½% mortgage note, payable in monthly payments of $3,964, secured by a building, due August 2010	**399,000**	—
Capitalized oil property purchase obligation	**59,000**	113,000
Legal settlement discounted at an imputed interest rate of 13.25% net of unamortized discount of $2,000 and $16,000, respectively, payable in semi-annual installments of $15,000, due March 1982	**28,000**	44,000
	1,128,000	4,913,000
Less amount included in current liabilities	**632,000**	1,993,000
	$ 496,000	$2,920,000

In May 1981, Basic renewed a financing agreement that allowed it to refinance its two bank notes. The renewal is structured as a revolving note in the amount of $10,000,000, payable in 48 monthly principal payments beginning April 1982 with interest payable monthly. The note bears interest at 1.5% above prime up to a maximum set by Texas usury laws, and is secured by certain producing oil and gas properties.

Under the loan agreement, Basic, without written permission from the bank, is restricted, among other things, from incurring additional indebtedness within certain defined limits, from making investments or capital expenditures in other than existing lines of business, or from selling any material part of its property. The maximum principal balance outstanding at any time is limited to 50% of the discounted present value of future operating cash flow from pledged oil and gas properties. Basic agreed to maintain current assets in excess of current liabilities (as defined), tangible net worth (as defined) of at least $4,000,000 and funded debt to tangible net worth of less than 1.1 to 1.

Total principal payments required under long-term debt agreements for the five years subsequent to March 31, 1981 are as follows:

Year ended March 31,	Amount
1982	$ 632,000
1983	—
1984	—
1985	—
1986	108,000
Thereafter	388,000
	$1,128,000

4. Employee Stock Ownership Plan

Basic adopted an Employee Stock Ownership Plan (ESOP) in March 1977. The ESOP was adopted effective April 1, 1976 and covers substantially all employees who have worked for Basic for six months and are employed at March 31. The contribution is determined by the Board of Directors annually and was 15% in 1980 of eligible employees' base salary which resulted in a contribution of $50,000 for the year ended March 31, 1980. The contribution was made in shares of Basic's common stock based upon the market price at March 31.

Basic did not make a contribution to the ESOP for the year ended March 31, 1981 because of tax law provisions that do not allow additions to an ESOP in excess of 25% of a participant's annual base pay. Forfeitures of shares from severed employees in the prior year are redistributed to the current plan participants. Forfeitures distributed at March 31, 1981 covered the entire 25%; therefore, Basic was prevented from making a contribution.

Notes to Consolidated Financial Statements

Basic Earth Science Systems, Inc.

5. Income Taxes

Basic and its subsidiaries file consolidated U.S. income tax returns and have elected to deduct intangible drilling and development costs in the period they are incurred. Basic reported tax losses of $2,938,000, $1,839,000 and $952,000 for the years ended March 31, 1981, 1980, and 1979, respectively. The benefit of the operating loss carryforwards has not been recognized because realization was not assured.

Net operating losses and investment tax credits available to offset taxable income and income taxes, respectively, in future years are as follows (see Note 6):

Year expiring	Net operating losses	Investment tax credits
1982	$ —	$ 8,000
1983	$ —	$ 23,000
1984	$ —	$ 10,000
1985	$ 323,000	$ 17,000
1986	$ 952,000	$ 4,000
1987	$1,839,000	$ 50,000
1988	$2,938,000	$148,000

In addition, Basic has a carryover of percentage depletion of $541,000 which is available to offset future taxable income (see Note 6). There is no limitation on the carryover period. Deferred income taxes have not been established on timing differences because of the availability of operating loss, percentage depletion and investment tax credit carryovers.

6. Contingencies

In April 1981, the Internal Revenue Service proposed adjustments to the consolidated income tax return for Basic for the year ended March 31, 1973, asserting that Basic understated taxable income of its construction subsidiaries. The nature of the understatement involved alleged nondeductible withdrawals by officers, directors and others (who have not been officers, directors or employees of Basic or its subsidiaries since June 1973) and alleged improper payments to government officials. Such alleged withdrawals and payments were made by the subsidiaries prior to the merger with Basic, but were deducted as construction costs on the consolidated return with Basic for the year ended March 31, 1973.

The Internal Revenue Service's proposal for tax deficiencies, penalties and interest could approximate $2,500,000 as of March 31, 1981 after utilization of available tax carrybacks and credits (see Note 5).

In management's opinion, there are a number of defenses against the proposal or certain portions of the proposal, and management intends to vigorously defend against the proposal. Basic has instituted legal action against the responsible individuals for recovery of any loss and damage arising from income tax deficiencies resulting from their acts. It is not possible to reasonably estimate the ultimate net liability, if any, that will accrue to Basic and, accordingly, no accrual has been made as of March 31, 1981.

Since June 9, 1981, the Company, as well as certain present and former officers and directors, has been sued in five separate class actions in the United States District Court, Southern District of New York. Basic's attorneys have been advised by plaintiffs' counsel that a consolidated amended complaint will be filed. The proposed consolidated action joins as defendants: John Muir & Co. (principal and managing underwriter for Basic's recent public offering of December 1980), a number of members of the Underwriting Syndicate, Resource Services International, Inc. (the independent petroleum engineering firm which estimated Basic's oil and gas reserves), *et al.* In summary, the actions are based on allegations of violations of the Securities Act of 1933 and 1934, of alleged misrepresentations in connection with Basic's public offering, allegations of improper use of proceeds of the public offering, of alleged misrepresentations of oil and gas reserves; and claiming damages resulting by reason thereof. Additionally, the Securities and Exchange Commission has requested the voluntary production of documents relating to estimates of Basic's oil and gas reserves in connection with an informal inquiry of Basic. Basic has agreed to provide those documents to the Commission.

Management believes that the prospectus did not contain materially false and misleading statements and did not fail to disclose material facts. Management believes that the allegations of the complaint are without merit and intends to vigorously defend its position against the suit. It is too early at this stage of the proceedings to give any opinion as to the outcome of this litigation or its possible effect on Basic.

7. Quasi-Reorganization

In September 1974, at the annual meeting of the shareholders of Basic, the shareholders approved a quasi-reorganization of Basic at April 1, 1974. This action resulted in the accumulated deficit at March 31, 1974 being charged against capital in excess of par value.

8. Related Party Transactions

During the year ended March 31, 1981, Basic entered into agreements with certain parties in which Basic is assigning to them a 5% working interest in the West Cole North Unit, Webb County, Texas. In consideration of the assignment, these parties are providing $412,000 to pay 12.5% of the estimated $3,300,000 cost of development of a full waterflood system on this unit. Basic is retaining a 92.93% working interest in the unit and will pay an estimated $2,818,000, 85.43% of the total estimated cost of the development. Included in the assignment was a 2% working interest to G. W. Breuer, president, director and a major shareholder.

Basic assigned to G. W. Breuer interests in certain prospects in Barber County, Kansas. Two wells drilled have been determined to be producing and are awaiting completion and two were dry holes. Mr. Breuer's investment was $49,000.

During the year ended March 31, 1980, Basic entered into a drilling arrangement in which G. W. Breuer was also a working interest owner. This exploration effort was not successful and was subsequently abandoned. Mr. Breuer's investment was approximately $100,000.

During the year ended March 31, 1979, Basic sold a 40% working interest in the West Cole Unit in Webb County, Texas. Under terms of the agreement, the new working interest owners funded a $1,500,000 water flood project. Any development costs and expenses exceeding $1,500,000 were to be shared proportionately by all working interest parties including Basic. Development funds advanced by the new working interest owners and not used were to be refunded to them on a pro rata basis. Basic estimated that $1,500,000 would be sufficient to complete the project. Basic had previously owned a 100% working interest in the property. Basic does not record gains or losses on transactions of this type because they are considered to be a pooling of assets in a joint undertaking to develop oil and gas. Mr. Breuer acquired a total of 25% of the working interest in this property. During the year ended March 31, 1980, Basic also sold lease and well equipment to these 40% working interest owners for $157,000. Basic's carrying value of this equipment was $99,000.

All of the above mentioned transactions with G. W. Breuer were on the same terms and conditions accepted by other independent third parties.

During the year ended March 31, 1981, Mr. Breuer assigned the remaining portion of his interest in the North Buffalo prospect in South Dakota to Basic. Basic's investment was approximately $131,000 in the prospect. The cost to Basic of this assignment was Mr. Breuer's cost plus 8%. This percentage over cost was similar to the portion of this acreage Mr. Breuer had previously assigned to an independent third party.

9. Income (Loss) Per Share

Earnings (loss) per share was determined by dividing the net income (loss) by the weighted average number of shares outstanding for each period.

10. Discontinued Operations

Petroleum services segment:

On January 17, 1980, the Board of Directors of Basic authorized the discontinuance of its petroleum services operations which were conducted by its wholly-owned subsidiaries, Beasley Trucking, Inc. and Tanks, Inc. On April 3, 1980, Basic sold a substantial portion of its petroleum services equipment at an auction.

The assets of this discontinued segment were reclassified at the date of discontinuance to remove them from their historical classifications and to separately identify them at their estimated net realizable value. The amount of assets and revenue applicable to the discontinued segment is presented to the right.

March 31,	**1981**	1980
Accounts receivable	**$ —**	$ 462,000
Assets held for resale	**175,000**	2,310,000
Other	**—**	54,000
Total assets at estimated net realizable value	**$175,000**	$2,826,000
Revenue — year ended March 31, 1980		$3,837,000

Notes to Consolidated Financial Statements
Basic Earth Science Systems, Inc.

10. Discontinued Operations (cont'd)

Consulting segment:

On March 1, 1979, Basic consummated an agreement to sell its entire stock holdings (425,000 shares) of the capital stock of Basic Earth Science Systems Indonesia, Inc. (Bessindo). The shares involved were 85% of the total 500,000 shares of Bessindo shares issued and outstanding. In settlement of an intercompany payable to Basic, Bessindo gave Basic an unsecured 54-month note for $135,000 (payable $2,500 per month) interest free, which Basic recorded on a discounted basis for $98,000. Bessindo assigned other assets to Basic, including a $64,000 note of the Buyer, which Bessindo formerly held. This note bears interest at 7% per year and is payable in monthly installments of $1,347 including interest, and is secured by the personal residence of the Buyer in Jakarta, Indonesia. Basic agreed to assume all U.S. liabilities of Bessindo in excess of $75,000.

Operations of the two discontinued segments are summarized below:

Year ended March 31,	1980	1979
Loss from operations of discontinued segments:		
Petroleum services, net of income tax recovery of $456,000 in 1980	$812,000	$2,104,000
Consulting	—	156,000
	$812,000	$2,260,000
Loss of disposal of segments:		
Petroleum services	$781,000	$ —
Consulting	—	98,000
	$781,000	$ 98,000

During the year ended March 31, 1980, Basic incurred a loss on the disposal of its petroleum services segment of $781,000, computed as follows:

Net carrying value of equipment sold	$1,957,000
Less proceeds of sale	1,863,000
Loss on revaluing equipment to net realizable value	94,000
Investment tax credit recapture	43,000
Loss from operations from January 17, 1980 to March 31, 1980	644,000
	$ 781,000

During the year ended March 31, 1979, Basic incurred a loss on the disposal of Bessindo of $98,000, computed as follows:

Investment in and intercompany receivable from Bessindo	$ 229,000
Proceeds from sale and value of net assets transferred in settlement of the intercompany payable to Basic	131,000
	$ 98,000

11. Change in Method of Accounting

During the year ended March 31, 1981, Basic changed its method of accounting for interest costs to conform to Statement on Financial Accounting Standards No. 34 entitled "Capitalization of Interest Cost." Basic's total interest cost for the year ended March 31, 1981 is as follows:

Total interest cost	$721,000
Amount capitalized as cost of property and equipment	175,000
Interest expense	$546,000

There is no effect on earnings per share because of this change in accounting principle.

12. Shareholders' Equity

In January 1981, Basic issued warrants to purchase 2,200,000 shares of common stock as part of its secondary offering. These warrants are exercisable at a price of $7.875 per share to June 17, 1982 and $9.00 per share from June 18, 1982 to June 17, 1983. No warrants have been exercised at March 31, 1981.

As part of the above secondary offering, Basic sold to the underwriter warrants to purchase 400,000 shares of common stock. These warrants can be exercised at a price of $9.00 per share from December 18, 1981 to December 17, 1985.

On April 28, 1981, the Board of Directors of Basic authorized a stock split of an additional share of stock for each share held of Basic's common stock to the shareholders of record on May 12, 1981. The issuance date of the additional shares of stock was May 26, 1981. Retroactive recognition of this stock split has been given in all financial statements and notes.

13. Reclassifications

Certain reclassifications have been made in the 1979 and 1980 financial statements in order to conform with the 1981 presentation.

14. Reserves of Oil and Gas — Unaudited

This footnote presents information relative to Basic's estimated oil and gas reserves and illustrates the estimated effect of such reserves on Basic's financial position and result of operations.

Information in this footnote is based on estimates of future production which may differ materially from actual recovery as production occurs.

Summary of Oil and Gas Operations Prepared on the Basis of Reserve Recognition Accounting (RRA)

Year ended March 31,	**1981**	1980
Additions and revisions to present value of estimated proved oil and gas reserves:		
Additions to estimated proved reserves, net of estimated future development costs	**$ 27,415,000**	$102,992,000
Revisions to estimates of reserves proved in prior years:		
Changes in price	**(6,800,000)**	9,226,000
Other	**(77,029,000)**	(7,839,000)
Accretion of discount	**10,603,000**	332,000
	(73,226,000)	1,719,000
Total additions and revisions to proved reserves	**(45,811,000)**	104,711,000
Acquisition, exploration and development costs incurred in connection with current period discoveries	**(2,080,000)**	(1,310,000)
Additions and revisions to proved reserves over evaluated costs	**(47,891,000)**	103,401,000
Recovery (provision) for income taxes (liability method)	**17,925,000**	(36,425,000)
Results of oil and gas producing activities on the basis of RRA	**(29,966,000)**	66,976,000
Other oil and gas operations net of income tax recovery of $506,000 at March 31, 1981 and $200,000 at March 31, 1980	**(593,000)**	(230,000)
Net income (loss) from oil and gas operations on the basis of RRA	**$(30,559,000)**	$ 66,746,000
Earnings (loss) per share on the basis of RRA	**$(2.93)**	$6.77
Weighted average number of outstanding shares	**10,432,600**	9,854,140

Notes to Consolidated Financial Statements

Basic Earth Science Systems, Inc.

14. Reserves of Oil and Gas — Unaudited (cont'd)

Summarized Consolidated Balance Sheet Prepared on the Basis of Reserve Recognition Accounting (RRA)

Assets

March 31,	**1981**	1980
Current assets	**$ 7,533,000**	$ 3,665,000
Other assets	**—**	96,000
Property and equipment:		
Present value of estimated future net revenues	**62,004,000**	106,030,000
Unevaluated acquisition cost	**2,752,000**	692,000
Other property and equipment, net	**807,000**	251,000
	$ 73,096,000	$110,734,000

Liabilities and Shareholders' Equity

Current liabilities	**$ 2,990,000**	$ 3,862,000
Long-term debt	**496,000**	2,920,000
Income taxes (liability method)	**17,338,000**	35,769,000
	20,824,000	42,551,000
Shareholders' equity:		
Retained earnings	**30,512,000**	61,071,000
Other	**21,760,000**	7,112,000
	52,272,000	68,183,000
	$ 73,096,000	$110,734,000

Analysis of Changes in Present Value of Estimated Future Net Revenue From Proved Oil and Gas Reserves

Year ended March 31,	**1981**	1980
Net present value at beginning of period	**$106,030,000**	$ 3,322,000
Revisions to reserves proved in prior years	**(73,226,000)**	1,719,000
Improved recovery techniques	**6,815,000**	16,658,000
New field discoveries and extensions	**20,600,000**	86,334,000
Net additions and revisions	**(45,811,000)**	104,711,000
Expenditures which reduced estimated future development costs	**3,884,000**	—
Sales of oil and gas, net of production costs of $1,180,000 and $647,000 at March 31, 1981 and 1980, respectively	**(2,099,000)**	(2,003,000)
Net present value at end of period	**$ 62,004,000**	$106,030,000

Notes to Consolidated Financial Summaries Prepared on the Basis of Reserve Recognition Accounting (RRA)

A. Reserve Recognition Accounting (RRA) Policies

The following accounting policies have been followed where RRA presentations differ from generally accepted accounting principles (GAAP). The estimates of proved reserves and related valuations were developed in accordance with the rules of the Securities and Exchange Commission (SEC) by Basic's independent petroleum engineers, G. J. Schneider at March 31, 1980 and Resource Services International, Inc. at March 31, 1981.

Under RRA, an asset is recognized and income recorded when proved reserves are added through exploration and development activities.

A dollar valuation (the RRA valuation) of proved reserves is developed as follows:

1. Estimates are made of quantities of proved reserves and the future periods during which they are expected to be produced based upon economic conditions that existed at the date of the reserve report.

2. The estimated future production of proved reserves is based on the current prices at the reserve report date except that future prices are increased for fixed and determinable escalation provisions (if any).

3. The resulting future gross revenue streams are reduced by estimated future costs to develop and produce the proved reserves, based on cost estimates at the date of the reserve report (estimated future net revenue).

14. Reserves of Oil and Gas — Unaudited (cont'd)

4. The resulting future net revenue streams are reduced to present value amounts by applying a 10% discount factor (present value of estimated future net revenues).

As acknowledged by the SEC, this valuation procedure does not necessarily yield the best estimate of the fair market value of a company's oil and gas properties. An estimate of fair market value should also take into account, among other factors, the likelihood of future recoveries of oil and gas in excess of proved reserves and anticipated future prices of oil and gas and related development and production costs. In the opinion of management, the estimated fair market value of Basic's oil and gas properties could be in excess of the RRA valuation of its proved reserves.

New field discoveries and extensions are included in earnings as they occur and represent proved reserves added from drilling exploratory and development wells. Subsequent revisions to the RRA valuation of proved reserves are included in earnings as they occur. The major factors causing revisions in proved reserves based on an annual RRA valuation are described below:

1. "Changes in Price" represents the approximate effect of increases or decreases from one period to the next in the price of oil and gas used in the RRA valuation calculation.

2. "Other" primarily represents the impact of the "Windfall Profit" tax, changes in engineering estimates of quantities and timing of production of reserves due to production history of the properties, additional development drilling, general market conditions, changes in production costs and various other considerations.

3. "Interest Factor — Accretion of Discount" represents the effect of the passage of time on the discounted value of proved reserves which is equal to 10% of the RRA valuation as of the beginning of the period.

Costs incurred during the period and expensed for RRA purposes include acquisition, exploration and development costs of properties evaluated in the current year. Costs of acquiring unproved properties and drilling exploration wells are deferred until the properties are evaluated, at which time they are charged to expense. All deferred costs are assessed for recoverability and any impairment is included in current expenses.

The RRA summary of operations for the year ended March 31, 1980 does not include Basic's discontinued petroleum services segment which incurred a loss of $1,593,000.

The provision for income taxes for oil and gas operations has been computed using the liability method. The estimated liability at the beginning and end of the period is computed by applying the statutory income tax rates at the balance sheet date to the difference between (a) the RRA valuation of proved reserves and (b) the tax basis of proved oil and gas properties after taking into consideration net operating loss and percentage depletion carryovers, and the present value of estimated statutory depletion and investment tax credits.

B. Estimated Future Net Revenues From Proved Reserves and Estimated Present Value of Proved Reserves

Estimated Future Net Revenues

As estimated at March 31, 1981

Year ending March 31,	Proved developed reserves	Proved undeveloped reserves	Total proved reserves
1982	$ 3,793,000	*$ 214,000	$ 4,007,000
1983	4,984,000	* 1,909,000	6,893,000
1984	5,151,000	* 3,061,000	8,212,000
Remainder	49,504,000	* 48,154,000	97,658,000
	$63,432,000	$53,338,000	$116,770,000

As estimated at March 31, 1980

Year ending March 31,	Proved developed reserves	Proved undeveloped reserves	Total proved reserves
1981	$ 4,690,000	**$ 2,722,000	$ 7,412,000
1982	5,811,000	** 14,021,000	19,832,000
1983	6,701,000	14,732,000	21,433,000
Remainder	51,481,000	82,007,000	133,488,000
	$68,683,000	$113,482,000	$182,165,000

*Net of estimated development costs of $1,832,000, $2,562,000, $3,433,000 and $426,000, respectively.

**Net of estimated development costs of $4,373,000 and $2,000,000, respectively.

Notes to Consolidated Financial Statements

Basic Earth Science Systems, Inc.

14. Reserves of Oil and Gas — Unaudited (cont'd)

Estimated Present Value of Proved Reserves

	Proved developed reserves	Proved undeveloped reserves	Total proved reserves
March 31, 1981	$33,651,000	$28,353,000	$ 62,004,000
March 31, 1980	$40,184,000	$65,846,000	$106,030,000

C. Proved Oil and Gas Reserve Quantities

	Oil and natural gas liquids (bbls.)	Natural gas (Mcf.)
Proved developed and undeveloped reserves at April 1, 1979	440,000	3,633,000
Revisions of previous estimates	246,000	213,000
Improved recovery	993,000	169,000
Discoveries	2,803,000	20,298,000
Production	(76,000)	(376,000)
Proved developed and undeveloped reserves at March 31, 1980	4,406,000	23,937,000
Revisions of previous estimates	(2,534,000)	(14,857,000)
Improved recovery	475,000	185,000
Discoveries	987,000	8,399,000
Production	(73,000)	(468,000)
Proved developed and undeveloped reserves at March 31, 1981	3,261,000	17,196,000

Proved reserves	Developed	Undeveloped	Total
Oil and natural gas liquids (bbls.)			
March 31, 1981	2,108,000	1,153,000	3,261,000
March 31, 1980	2,227,000	2,179,000	4,406,000
Gas (Mcf.)			
March 31, 1981	5,151,000	12,045,000	17,196,000
March 31, 1980	5,264,000	18,673,000	23,937,000

D. Income Taxes

The income tax provision was estimated assuming the utilization of existing operating losses, percentage depletion carryovers and the present value of estimated future percentage depletion and investment tax credits.

The following is a reconciliation of the income tax liability to the 46% statutory rate:

March 31,	**1981**	1980
Tax on estimated present value in excess of tax basis	**$25,740,000**	$47,071,000
Tax effect at 46% of:		
Operating loss and percentage depletion carryforwards	**(3,364,000)**	(1,738,000)
Present value of estimated future percentage depletion	**(4,651,000)**	(9,221,000)
Present value of investment tax credits (including estimated investment tax credits on future development costs)	**(387,000)**	(343,000)
	$17,338,000	$35,769,000

14. Reserves of Oil and Gas — Unaudited (cont'd)

This computation does not take into consideration the effects of other income and expense of Basic which will be incurred in the future which may have a material effect on income taxes and the realization of operating loss and investment tax credit carryovers and future percentage depletion.

The tax liability disclosed above is classified as long-term and no income taxes are currently payable.

E. Management's Comments on RRA Presentation

As disclosed in the Summary of Oil and Gas Operations prepared on the basis of Reserve Recognition Accounting, reserve values increased by $27,415,000. Of this increase, approximately 25% resulted from improved recovery techniques on the West Cole North Unit. The remaining 75% resulted from Basic's increased exploration program which resulted in approximately eight net exploratory wells during the year ended March 31, 1981. Reserve values decreased during the year ended March 31, 1981 by $73,226,000 because of other revisions of prior estimates. Approximately 65% of these revisions resulted from information obtained from dry development wells drilled in the Buck Creek Field and in the Playa Flank area. Additionally, revised estimates of proved developed reserves on the Hilton leases and the West Cole Unit accounted for approximately 25% of this change. These changes resulted primarily from additional production history available on these properties.

The estimated present value of proved reserves as computed under SEC guidelines may not necessarily represent the fair value of Basic's oil and gas properties in the market-place. Other factors such as increasing prices and costs and improved recovery techniques may have significant effects upon the amount of recoverable reserves and their present value. Additionally, future net revenues from proved undeveloped reserves, as shown in the preceding tables, are net of estimated future costs to develop the reserves for production. Basic's estimate of such costs (approximately $8,253,000 and $6,373,500 at March 31, 1981 and 1980, respectively) contemplated development of the reserves and incurrence of the costs over a four-year period.

Fluctuations of interest rates and the possibility of additional reserves underlying Basic's undeveloped leases may materially affect the fair value. Undeveloped leases in the consolidated balance sheet on the basis of Reserve Recognition Accounting are carried at cost, less a provision for impairment, while such leases may have value in excess of cost.

Disclosures of reserves do not include any "probable" and "possible" reserves which may exist underlying Basic's leases and may become available through additional development drilling activity. No value has been assigned to such reserves in the consolidated balance sheet.

Because earnings under RRA are recognized upon discovery of proved reserves and when the RRA valuation of proved reserves change, no earnings are reported as oil and gas are produced. Consequently, RRA earnings will differ substantially from funds generated or required by current exploration, development and producing operations. For information on funds flow, see the Consolidated Statement of Changes in Financial Position.

Report of Certified Public Accountants

The Board of Directors and Shareholders
Basic Earth Science Systems, Inc.
Denver, Colorado

We have examined the consolidated balance sheets of Basic Earth Science Systems, inc. at March 31, 1981 and 1980 and the related consolidated statements of operations, changes in shareholders' equity and changes in financial position for each of the three years in the period ended March 31, 1981. Our examinations were made in accordance with generally accepted auditing standards and, accordingly, included such tests of the accounting records and such other auditing procedures as we considered necessary in the circumstances.

As discussed in Note 6, the Internal Revenue Service has asserted a claim for the understatement of taxable income against the Company. Additionally, the Company is a defendant in a number of class action lawsuits brought by some of Basic's shareholders. The Company intends to vigorously defend against these actions. It is not possible, at this time, to reasonably estimate the ultimate outcome of these actions.

In our opinion, subject to the effects of such adjustments, if any, as might have been required had the outcome of the uncertainties referred to in the preceding paragraph been known, the financial statements designated above present fairly the financial position of Basic Earth Science Systems, Inc. at March 31, 1981 and 1980 and the consolidated results of its operations and changes in its shareholders' equity and financial position for each of the three years in the period ended March 31, 1981, in conformity with generally accepted accounting principles which, except for the change with which we concur, in the method of recording interest as described in Note 11, have been applied on a consistent basis.

Fox & Company

Denver, Colorado
June 10, 1981

Part II — Amarex, Inc.

Consolidated Statements of Income

Amarex, Inc. and Subsidiaries

	Year Ended April 30		
	1981	1980	1979
Revenues			
Gas sales	$19,885,000	$11,655,000	$ 8,572,000
Oil sales	6,607,000	5,064,000	3,315,000
Other	2,185,000	367,000	461,000
	28,677,000	17,086,000	12,348,000
Costs and Expenses			
Operating expenses	4,293,000	2,543,000	2,123,000
Windfall profit tax	1,076,000	156,000	
General and administrative	3,063,000	1,807,000	1,487,000
Depreciation and depletion	6,101,000	3,287,000	3,273,000
Interest	9,657,000	5,110,000	1,983,000
	24,190,000	12,903,000	8,866,000
Income Before Income Taxes	4,487,000	4,183,000	3,482,000
Provision for income taxes	1,070,000	1,274,000	964,000
Net Income	$ 3,417,000	$ 2,909,000	$ 2,518,000
Net income per common share, based on average shares outstanding	$ .41	$ .36	$ .31

See notes to consolidated financial statements.

Consolidated Balance Sheets

Amarex, Inc. and Subsidiaries

30

Assets

	April 30	
	1981	1980
Current Assets		
Cash	**$ 3,964,000**	$ 2,525,000
Short-term investments, at cost which approximates market	**7,912,000**	
Receivables		
Trade accounts	**22,218,000**	6,514,000
Drilling partnerships	**5,499,000**	6,983,000
Inventories	**4,327,000**	1,396,000
Prepaid expenses	**510,000**	136,000
Total Current Assets	**44,430,000**	17,554,000
Short-Term Investments Held for Property Additions	**14,401,000**	
Other Assets	**6,670,000**	1,469,000
Property and Equipment		
Gas and oil properties — Full cost method of accounting	**156,882,000**	80,251,000
Other property and equipment	**4,698,000**	2,150,000
	161,580,000	82,401,000
Less allowance for depreciation and depletion	**(22,358,000)**	(16,249,000)
	139,222,000	66,152,000
	$204,723,000	$85,175,000

Liabilities and Stockholders' Equity

	April 30	
	1981	1980
Current Liabilities		
Trade accounts payable	**$ 21,048,000**	$12,249,000
Accrued expenses and other current liabilities	**8,593,000**	3,810,000
Current maturities of long-term debt	**261,000**	202,000
Total Current Liabilities	**29,902,000**	16,261,000
Deferred Revenue	**452,000**	344,000
Deferred Income Taxes	**4,186,000**	3,120,000
Long-Term Debt	**114,222,000**	47,113,000
Leases and Contingencies – Note G		
Stockholders' Equity		
Common stock, no par value:		
Authorized shares, 1981 - 50,000,000; 1980 - 5,000,000; shares issued and outstanding, 1981 - 9,252,906; shares issued, 1980 - 4,224,095; outstanding, 1980 - 4,035,688	**43,686,000**	4,224,000
Paid-in capital		6,187,000
Retained earnings	**12,275,000**	9,826,000
Treasury stock; 188,407 shares in 1980		(1,900,000)
	55,961,000	18,337,000
	$204,723,000	$85,175,000

See notes to consolidated financial statements.

Consolidated Statements of Stockholders' Equity

Amarex, Inc. and Subsidiaries

32

	Common Stock	Paid-In Capital	Retained Earnings	Treasury Stock
Balance at April 30, 1978	$ 2,172,000	$ 7,917,000	$ 4,399,000	$(1,900,000)
Common Stock issued upon exercise of stock options (25,800 shares)	26,000	143,000		
Net income			2,518,000	
Balance at April 30, 1979	2,198,000	8,060,000	6,917,000	(1,900,000)
Common Stock issued upon exercise of stock options (14,700 shares)	15,000	138,000		
Two-for-one stock split (2,011,344 shares)	2,011,000	(2,011,000)		
Net income			2,909,000	
Balance at April 30, 1980	4,224,000	6,187,000	9,826,000	(1,900,000)
Common Stock issued upon exercise of stock options (9,430 shares)	46,000	44,000		
Change to no par value stock . . .	6,231,000	(6,231,000)		
Retirement of 188,407 shares held in the Treasury	(932,000)		(968,000)	1,900,000
Two-for-one stock split (4,042,788 shares)				
Issuance of Common Stock in public offering (1,165,000 shares)	34,117,000			
Net income			3,417,000	
Balance at April 30, 1981	$43,686,000	$ —	$12,275,000	$ —

See notes to consolidated financial statements.

Consolidated Statements of Changes in Financial Position

Amarex, Inc. and Subsidiaries

	Year Ended April 30		
	1981	1980	1979
Sources of Funds			
From Operations:			
Net income	**$ 3,417,000**	$ 2,909,000	$ 2,518,000
Add items not affecting working capital:			
Depreciation and depletion	**6,101,000**	3,287,000	3,273,000
Deferred income taxes	**1,066,000**	1,258,000	934,000
Total From Operations	**10,584,000**	7,454,000	6,725,000
Increase in deferred revenue	**108,000**	5,000	36,000
Proceeds from long-term borrowing	**164,126,000**	24,943,000	13,618,000
Proceeds from sales of property and equipment	**825,000**	854,000	1,608,000
Issuance of Common Stock	**34,207,000**	153,000	169,000
Total Sources of Funds	**$209,850,000**	$ 33,409,000	$ 22,156,000
Uses of Funds			
Additions to property and equipment	**$ 79,996,000**	$ 34,805,000	$ 15,463,000
Payments of long-term debt and change in current maturities	**97,017,000**	337,000	9,011,000
Increase (decrease) in short-term investments held for property additions and other assets	**19,602,000**	(17,000)	(299,000)
Increase (decrease) in working capital	**13,235,000**	(1,716,000)	(2,019,000)
Total Uses of Funds	**$209,850,000**	$ 33,409,000	$ 22,156,000
Working Capital			
Increase (decrease) in current assets:			
Cash	**$ 1,439,000**	$ (146,000)	$ 113,000
Short-term investments	**7,912,000**	—	—
Trade accounts receivable	**15,704,000**	1,716,000	263,000
Drilling partnerships receivable	**(1,484,000)**	6,152,000	(1,352,000)
Inventories	**2,931,000**	953,000	(274,000)
Prepaid expenses	**374,000**	78,000	(105,000)
	26,876,000	8,753,000	(1,355,000)
(Increase) decrease in current liabilities:			
Trade accounts payable	**(8,799,000)**	(8,479,000)	(619,000)
Accrued expenses and other current liabilities	**(4,783,000)**	(1,930,000)	(79,000)
Current maturities of long-term debt	**(59,000)**	(60,000)	34,000
	(13,641,000)	(10,469,000)	(664,000)
Increase (decrease) in working capital	**$ 13,235,000**	$(1,716,000)	$(2,019,000)

See notes to consolidated financial statements.

Notes to Consolidated Financial Statements

April 30, 1981

Amarex, Inc. and Subsidiaries

Note A – SUMMARY OF SIGNIFICANT ACCOUNTING POLICIES

Principles of Consolidation

The consolidated financial statements include the accounts of the Company and its wholly-owned domestic subsidiaries. The investment in a foreign subsidiary (included in gas and oil properties) is carried at cost which approximates the Company's investment determined by the equity method. All significant intercompany accounts and transactions have been eliminated upon consolidation.

The Company follows the proportional consolidation method of accounting for its investment in drilling partnerships. The Company's pro rata share of the assets, liabilities and operations of such drilling partnerships is included in the consolidated financial statements.

Inventories

Inventories consist primarily of tubular goods valued at the lower of cost (specific identification) or market.

Property and Equipment

The Company follows the full cost method of accounting for gas and oil properties. Accordingly, all costs associated with property acquisition, exploration and development of gas and oil reserves are capitalized.

The capitalized costs of gas and oil properties are amortized on a composite method based on estimated future revenues over the aggregate productive life of all gas and oil properties. Such costs include amounts for lease acquisition, producing properties, properties in progress, investments in and certain costs of organizing drilling partnerships, and, for purposes of amortization, an estimate for future development costs of proved reserves. The Company's gas and oil reserves are estimated annually by independent petroleum engineers. Future revenues from gas and oil reserves have been estimated based on rules prescribed by the Securities and Exchange Commission (SEC). Therefore, crude oil is priced at current selling prices including the effect of decontrol of crude oil prices under Department of Energy regulations net of the effect of the windfall profit tax. Natural gas is priced in accordance with the Natural Gas Policy Act (NGPA), using current prices adjusted for contractual increases, other fixed and determinable escalations under provisions of the NGPA, and the effect of applying current deregulated prices to production expected to occur after the deregulation dates stipulated by the NGPA if existing contracts permit price redeterminations. Depreciation and depletion per dollar of gas and oil sales was \$.23, \$.20, and \$.28 for the years ended April 30, 1981, 1980, and 1979, respectively.

No gains or losses are recognized upon the sale or disposition of producing gas and oil properties, except where a significant amount of reserves is involved; rather, the proceeds from sales are credited to producing gas and oil properties.

Depreciation of other property and equipment is provided using the straight-line method over estimated useful lives of 5 to 10 years.

Federal Income Taxes

Deferred federal income taxes are provided for all significant timing differences between financial and tax reporting, including certain gas and oil exploration and development costs that are capitalized and amortized for financial reporting purposes and expensed for federal income tax purposes. Other timing differences arise when income or expenses are recognized for financial reporting purposes in different periods than for income tax purposes. Investment tax credits are included in income in the year the qualified assets are placed in service, to the extent they are allowable for tax purposes or would have been allowable in the absence of timing differences.

Stock Options

Stock options were accounted for, when exercised, by a credit to capital stock and paid-in capital through September 18, 1980 and are accounted for by a credit only to capital stock after such date (See Note H). No charges or credits to income are recognized in connection with the options.

Debt Issue Expense

Debt issue expense, included in other assets at April 30, 1981, is being amortized by the interest method.

Short-term Investments

Short-term investments are carried at cost, which approximates market.

Net Income Per Common Share

Net income per common share is computed based on the weighted average number of common and common equivalent shares outstanding during the respective years. The effect of common stock equivalents on earnings per common share is not material.

Reclassifications

Certain reclassifications have been made in the 1980 and 1979 financial statements to conform to the classifications used in 1981.

Note B – SHORT-TERM INVESTMENTS

Short-term investments consist of certificates of deposit held by the trustee for the 12% non-recourse secured notes (See Note F). These investments are restricted to the payment of productive well intangible costs and productive well lease costs incurred by three limited partnerships in which the Company is a general partner At April 30, 1981, current assets include $7,912,000 which is restricted to pay trade accounts payable arising from such productive well costs.

Note C – GAS AND OIL PRODUCING ACTIVITIES

All of the Company's gas and oil producing activities have been conducted in the United States. Total costs incurred (whether capitalized or expensed) and depreciation and depletion in connection with the Company's gas and oil producing activities were as follows:

	Year Ended April 30		
	1981	1980	1979
Property acquisition	**$12,442,000**	$ 2,959,000	$ 2,462,000
Exploration	**44,191,000**	19,840,000	2,613,000
Development	**20,795,000**	11,381,000	10,356,000
Production	**4,549,000**	2,699,000	2,123,000
	$81,977,000	$36,879,000	$17,554,000
Provision for depreciation and depletion	**$ 5,941,000**	$ 3,250,000	$ 3,241,000

At the dates shown below, aggregate capitalized cost and related allowances for depreciation and depletion for gas and oil activities were as follows:

	April 30	
	1981	1980
Aggregate capitalized costs of proved and unproved properties	**$156,882,000**	$80,251,000
Allowance for depreciation and depletion	**(22,022,000)**	(16,081,000)
	$134,860,000	$64,170,000

The Company conducts a significant part of its operations through limited partnerships in which it is a general partner. Gas and oil properties include $81,162,000 and $29,906,000 at April 30, 1981 and 1980, respectively, invested directly in these partnerships. Included in gas and oil sales, from the Company's investment in drilling partnerships are $13,386,000, $8,922,000 and $7,243,000 for the years ended April 30, 1981, 1980 and 1979, respectively. Related amounts included in lease operations and general and administrative expenses are $2,807,000 in 1981, $1,673,000 in 1980 and $1,556,000 in 1979. Net revenues from production (gross revenues less lifting costs and windfall profit tax) for the years ended April 30, 1981, 1980 and 1979 were $21,943,000, $14,020,000 and $9,764,000, respectively.

Note D – ACCRUED EXPENSES AND OTHER CURRENT LIABILITIES

Accrued expenses and other current liabilities consist of the following:

	April 30	
	1981	1980
Funds held for other working interest owners	$1,148,000	$ 817,000
Accrued interest	1,830,000	952,000
Accrued employee compensation	952,000	628,000
Accrued partnership formation costs	3,588,000	620,000
Miscellaneous	1,075,000	793,000
	$8,593,000	$3,810,000

Note E – FEDERAL INCOME TAXES

A reconciliation of statutory federal income tax to the Company's effective federal income tax provision is as follows:

	Year Ended April 30		
	1981	1980	1979
	(in thousands)		
Federal tax based on statutory rate	$2,064	$1,924	$1,672
Decrease in taxes resulting from:			
Statutory depletion in excess of cost depletion	(598)	(423)	(544)
Utilization of investment tax credit	(400)	(236)	(180)
Other	-	(7)	(14)
Provision for federal income taxes	$1,066	$1,258	$ 934

In addition to the federal tax provision shown above, the total provision for income taxes includes state income taxes, all of which were currently payable, of $4,000, $16,000 and $30,000 for the years ended April 30, 1981, 1980, and 1979, respectively.

Deferred taxes result principally from certain costs related to producing gas and oil properties which are expensed currently for tax purposes and capitalized in gas and oil properties for financial reporting purposes. Deferred taxes have been provided to the extent that net timing differences exceed tax depletion carryforward and the effect of utilizing investment tax credits.

The components of deferred tax expense are as follows:

	Year Ended April 30		
	1981	1980	1979
	(in thousands)		
Exploration and development costs expensed for tax purposes and capitalized for financial reporting purposes	$20,579	$9,112	$3,700
Timing differences in excess of amounts necessary to eliminate current tax provisions	(18,600)	(7,560)	(2,048)
Statutory depletion in excess of cost depletion	(598)	(423)	(544)
Utilization of investment tax credit	(400)	(236)	(180)
Partnership income recognized for financial statement purposes in excess of amounts for tax purposes	85	365	6
	$ 1,066	$1,258	$ 934

At April 30, 1981, the Company had a net operating loss carryover of $63,863,000 and an investment tax credit carryover of $1,416,000, which will expire, if not utilized, as follows:

Year of Expiration	Net Operating Loss	Investment Tax Credit
1982	$ –	$ 209,000
1983	930,000	171,000
1984	–	50,000
1985	1,680,000	107,000
1986	4,267,000	180,000
1987	16,551,000	299,000
1988	40,435,000	400,000
	$63,863,000	$1,416,000

The Company has a tax depletion carryover of approximately $6,467,000 which is available indefinitely and can be utilized for tax purposes after utilization of the net operating loss carryover

The net operating loss carryover consists of timing differences between financial and tax reporting. Depletion and investment tax credit carryovers are differences for which the tax effects have been recognized for financial reporting purposes. Therefore, federal income tax expense for financial reporting will not be reduced for the periods in which such carryforwards are utilized for tax purposes.

The Internal Revenue Service has examined the Company's federal income tax returns through April 30, 1975. A substantial portion of the proposed adjustments has tentatively been successfully resolved. Management, on advice of tax counsel, plans to protest the remaining unresolved issues and believes they can be resolved without a material effect on the financial position of the Company. However, the net operating loss carryforwards expiring in 1983 and 1985 may be substantially less if the issues are not resolved in favor of the Company.

Note F – LONG-TERM DEBT

Long-term debt is summarized as follows:

	April 30	
	1981	1980
Senior debt:		
12% non-recourse secured notes due December 31, 1995 (1)	$ 45,134,000	$ –
Note payable to banks, interest at ½% above the bank prime rate (19% at April 30, 1980), (2)	–	38,650,000
Note payable to a bank, due October 1, 1985 (3)	16,150,000	–
Other long-term obligations due periodically through 1987 (4)	3,051,000	8,665,000
Subordinated debt:		
8¾% convertible subordinated oil and gas revenue debentures, due 1990 (5)	21,400,000	–
13¾% subordinated debentures, due April 1, 2000 - net of unamortized debt discount (6)	28,748,000	–
	114,483,000	47,315,000
Current maturities	(261,000)	(202,000)
	$114,222,000	$47,113,000

(1) In June 1980, the Company entered into agreements for the sale of $60,500,000 of 12% non-recourse secured notes due December 31, 1995, together with certain net profits royalty interests. The proceeds from the notes are held in trust until used to pay productive well intangible costs and productive well lease costs incurred by three limited partnerships (the "Programs") in which the Company is the general partner. The notes will be secured and repaid as to principal and interest only from 75% of the Company's interest in the net revenues generated by the Programs except that the Company has guaranteed payment of the first three years' interest on the notes. The net

profits royalty interest, which commences only after the notes, including all interest, have been repaid in full, provides the noteholders with a continuing interest of 10% of the net revenues of the Programs, payable out of the Company's interest therein. The Company has agreed to offer to repurchase up to 20% of the aggregate outstanding principal amount of the notes if at December 31, 1985 the present value of estimated future net revenues from certain proved reserves attributable to 75% of the Company's interest in the specified Programs is less than 120% of the then outstanding principal amount of the notes. If the Company is required to make the repurchase offer, those noteholders who accept the repurchase offer would receive an 8% net profits royalty interest rather than the 10% interest discussed above.

The noteholders have advanced three payments totaling $45,375,000 to the Trustee through April 30, 1981. Under the terms of the note agreements, the Trustee will release funds to reimburse the Company for productive well intangible costs and productive well lease costs incurred. Of the amount advanced by the noteholders, the Company has received $23,062,000, which represents substantially all of such costs incurred to April 30, 1981 by the Programs.

(2) A revolving credit agreement with a group of banks pledged the Company's interest in substantially all proven developed gas and oil reserves as security for advances under the agreement, except for such reserves pledged under the non-recourse note agreements. During April 1981, this note was fully paid and the agreement has since been terminated.

(3) Through April 1981, the Company was advanced $16,150,000 under a $20,000,000 line of credit agreement with a bank unrelated to the revolving credit agreement. The interest rate is the lower of 1½% above the rate paid on funds invested in certificates of deposits in a participating bank or the bank's prime rate, which was 18% at April 30, 1981. The line of credit permits advances in amounts up to the carrying value of undeveloped leases. Principal payments will not be required until October 1, 1985, so long as the carrying value of undeveloped leases are at least equal to total amounts advanced. Undeveloped leases with a carrying value of $16,861,000 are available to be pledged as security on this agreement at April 30, 1981.

(4) The Company has short-term lines of credit with three banks totaling $15,000,000. At April 30, 1981 a balance of $1,554,000 is outstanding under such agreements and the maturity date has been extended to May 1, 1982.

The Company has various other long-term obligations of $1,497,000 maturing at various dates or in installments through 1987 of which $375,000 is secured by a first lien on a gas gathering facility.

(5) In December 1980, the Company sold $21,400,000 principal amount of 8¾% Convertible Subordinated Oil and Gas Revenue Debentures due 1990 to investors outside of the United States. The debentures obligate the Company to expend a defined amount over a three-year period to develop specified prospects owned by the Company. At the end of the three-year exploration period, the holder of the Debentures may convert them into a specified royalty interest in the specified prospects. If not converted, the Debentures will mature in 1990, but may be retired at the option of the Company prior to maturity at the face amount plus accrued interest.

(6) In July 1980, the Company issued $30,000,000 principal amount of 13¾% unsecured subordinated debentures. The debentures are redeemable at the Company's option, in whole or in part, at an initial redemption price of 109.167% of the principal amount beginning August 1, 1985, at declining prices thereafter through July 1995, and thereafter at 100%. Commencing on August 1, 1990 and annually thereafter, through a sinking fund, the Company is required to redeem 7.5% of the original principal amount of debentures, subject to credit for debentures otherwise acquired by the Company. In addition, the Company has the right to make an additional sinking fund payment in an amount not exceeding the mandatory sinking fund payment for that year. The debentures are subordinated in right of payment to all existing and future senior indebtedness, as defined, of the Company. The debenture indenture limits payment of dividends and the purchase, redemption or retirement of the Company's capital stock.

Estimated annual principal payments under the terms of all the loans, including estimated amounts to be paid from future revenues, through 1986 are $8,004,000, $11,012,000, $16,449,000, $21,224,000 and $6,577,000.

Note G – LEASES AND CONTINGENCIES

The Company has various leases and renewal options, primarily for office space, requiring future rentals varying from $175,000 to $320,000 per year through April 30, 1984. Existing leases for office space expire on December 31, 1984; however, the Company expects that the leases will be renewed or that new leases will be negotiated. However, the above amounts are not necessarily indicative of future rental payments. During the years ended April 30, 1981, 1980 and 1979 net rental payments charged to expense amounted to $339,000, $132,000 and $115,000, respectively. There are no material capital leases.

The Company is contingently liable, as general partner of various partnerships, for notes payable to banks of $3,787,000 at April 30, 1981. The notes are secured by and will be paid from the other partners' share of the partnerships' revenues. The Company is also contingently liable as guarantor on the borrowing of a foreign subsidiary up to a maximum of $500,000.

As general partner in certain partnerships, the Company is obligated to make a repurchase offer according to the terms of the respective partnership agreements. The price for such an offer would be based on the fair market value of other assets and on reserves of the partnership as established by an independent engineer.

Natural gas sales are subject to regulation by the Federal Energy Regulatory Commission (FERC) under the Natural Gas Act and the Natural Gas Policy Act of 1978 (NGPA). The Company has filed for and is currently receiving the maximum allowable prices under FERC regulations. The Company has yet to receive final FERC approval on certain of the filings and, accordingly, a portion of the increased amounts being collected by the Company is subject to possible refund. In the opinion of management, such amounts recognized as sales are not material.

The maximum prices established by the NGPA for various categories of gas are intended to be ceiling prices and may be charged and collected only if contractually authorized. Gas sales contracts entered into prior to the enactment of the NGPA typically included "area rate" clauses. Such clauses vary among contracts, but typical clauses provide for price escalations whenever FERC or its predecessor establishes a higher applicable just and reasonable rate. The current policy of FERC is that such clauses will allow the maximum NGPA prices without contract amendments if both the purchaser and seller agree that such is their mutual interpretation of such clauses.

The current FERC policy is still subject to administrative and judicial review and, accordingly, a portion of the increased amounts being collected by the Company under gas contracts with such clauses could eventually be subject to a possible refund. However, until the administrative and judicial reviews are finalized no estimate can be made as to the amount of such a potential refund.

Note H – CAPITAL STOCK

The Company's Articles of Incorporation were amended on September 18, 1980 to increase the number of authorized shares of Common Stock from 5,000,000 to 50,000,000 and to change the par value of the Common Stock from $1 per share to no par value per share. Accordingly, all paid-in capital was transferred to Common Stock during September 1980.

On November 12, 1980, the Board of Directors authorized the retirement of the 188,407 shares of Common Stock of the Company held as treasury stock. Accordingly, retained earnings and Common Stock were reduced by $968,000 and $932,000, respectively.

On November 12, 1980, the Board of Directors authorized a two-for-one stock split on the outstanding shares of Common Stock of the Company to be distributed to shareholders of record at the close of business on November 30, 1980. The distribution of such shares was made on December 15, 1980. All share and per-share information (except for the historical stockholders' equity information) has been adjusted for this stock split. The effect of the stock split was to change earnings per share as follows:

	Year Ended April 30		
	1981	1980	1979
Earnings per share:			
Before split	$.82	$.72	$.63
After split	$.41	$.36	$.31

In March 1981, the Company completed a public offering of 1,165,000 shares of Common Stock at $31 per share. Proceeds from the offering, net of issue cost, were used to retire bank debt.

At April 30, 1981 and April 30, 1980, there were 1,000,000 shares of $1 par value Preferred Stock authorized but unissued.

Note I – STOCK OPTIONS

The following table reflects the shares of the Company's Common Stock reserved for issuance at April 30, 1981 and 1980, and has been adjusted to reflect the two-for-one stock split in November, 1980.

	Exercise Price Per Share	Expiration Date	Number of Shares Reserved for Issuance 1981	1980
Outstanding options	$4.03-$16.13	April 1982 through June 1985	37,870	44,400
Available for future grant			254,400	264,400
			292,270	308,800

The Qualified Stock Option Plans authorize the granting of options to officers and key employees to purchase 292,270 shares of Common Stock at a price not less than the market value on the date of the grant. The options outstanding at April 30, 1981 were granted as follows: 8,000, 2,200, 19,670 and 8,000 granted during the fiscal years ended April 30, 1977, April 30, 1978, April 30, 1980 and April 30, 1981, respectively. No options were granted under the 1969 Qualified Stock Option Plan after its expiration date, October 2, 1979. No options will be granted under the 1971 Qualified Stock Option Plan after its expiration date, July 15, 1981. Options granted under these two plans can be exercised 20% annually during a five-year period from the date of grant.

A non-qualified option plan authorizes the granting of options to officers and key employees to purchase 134,400 shares of Common Stock at prices not less than the market value on the date of grant. The options can be exercised 20% annually during a six-year period from the date of grant, but cannot be exercised until after six months from the date of grant.

At April 30, 1981 and 1980, 15,870 and 20,400 shares, respectively, were exercisable under the above plans.

The number of shares for which options were exercised and the related option price ranges, after giving effect to the two-for-one stock split in 1981 and 1980, under the Qualified Plan during the last three years were; 1981 16,530 shares at $1.91 to $16.13; 1980 - 32,800 shares at $4.03 to $4.75; 1979 - 3,200 shares at $2.03. Options were exercised under the Non-Qualified Plan in 1979 for 100,000 shares at $1.63. No options were exercised under the Non-Qualified Plan in 1980 and 1981.

Note J – BUSINESS SEGMENT AND MAJOR PURCHASERS

The Company is engaged in one line of business - the exploration for and acquisition, development, and production of gas and oil in the United States. During 1979 and 1980, the Company had revenues from Arkansas Louisiana Gas Company which accounted for more than 10% of gas and oil sales. During 1981, El Paso Natural Gas Company and Oklahoma Natural Gas Company became the second and third major customers by accounting for 24.9% and 12.0% of total sales, respectively. No other customer individually accounted for more than 10% of total sales to customers. Due to the high demand for gas and oil, the Company believes that should it lose any of its major customers such customers could be replaced without a material adverse effect on the Company.

Sales to major customers are summarized as follows:

	Year Ended April 30 1981	1980	1979
	(in thousands)		
Arkansas Louisiana Gas Company	$4,887	$4,870	$3,686
Percent of sales	18.4%	29.1%	31.0%
El Paso Natural Gas Company	$6,604	–	–
Percent of sales	24.9%	–	–
Oklahoma Natural Gas Company	$3,187	–	–
Percent of sales	12.0%	–	–

Note K – QUARTERLY FINANCIAL DATA (UNAUDITED)

Revenues, income before income taxes, net income, and earnings per share by quarter for the years ended April 30, 1981, and 1980 are presented below. All earnings per share amounts have been adjusted for the stock split as discussed in Note H.

	Year Ended April 30, 1981			
	Quarter			
	First	Second	Third	Fourth
	(in thousands except per share amount)			
Revenues	$5,778	$6,719	$7,030	$9,150
Income before income taxes	$ 903	$ 921	$ 923	$1,740
Net income	$ 604	$ 667	$ 720	$1,426
Earnings per share	$ .08	$ .08	$ .09	$ .16

	Year Ended April 30, 1980			
	Quarter			
	First	Second	Third	Fourth
	(in thousands except per share amount)			
Revenues	$3,541	$3,455	$4,650	$5,440
Income before income taxes	$ 981	$ 684	$1,013	$1,505
Net income	$ 708	$ 492	$ 747	$ 962
Earnings per share	$ .09	$ .06	$ .09	$ .12

Note L – SUPPLEMENTAL RESERVE INFORMATION (UNAUDITED)

The following information summarizes the Company's net proved reserves of gas and oil (including condensate) and the present values thereof for the years ended April 30, 1981, 1980 and 1979. The reserve estimates are based upon the report of Keplinger & Associates, Inc., independent petroleum engineers. Such estimates are in accordance with regulations prescribed by the Securities and Exchange Commission. Accordingly, crude oil is priced at current selling prices including the effect of decontrol of crude oil prices under Department of Energy regulations net of the effect of the windfall profit tax. Natural gas is priced in accordance with the Natural Gas Policy Act (NGPA), using current prices adjusted for contractual increases, other fixed and determinable escalations under provisions of the NGPA, and the effect of applying current deregulated prices to production expected to occur after the deregulation dates stipulated by the NGPA if existing contracts permit price redeterminations.

The reliability of any reserve estimate is a function of the quality of available information and of engineering interpretation and judgment. In management's opinion, the reserve estimates presented herein, in accordance with generally accepted engineering and evaluation principles consistently applied, are believed to be reasonable. These reserves should be accepted with the understanding that drilling activities or additional information subsequent to the date of this report might require their revision. Moreover, substantial quantities of the reserves included in this study were estimated using the volumetric method, and such estimates are particularly susceptible to revision in the light of subsequent drilling activities or information.

The valuation of proved reserves does not, in the Company's opinion, reflect expected actual revenues to be derived from those reserves nor their market value. In the Company's opinion, the valuation of proved reserves would include, among other things, factors such as (i) anticipated future increases of gas and oil prices and production and development costs; (ii) discount rates reflecting the costs of financing and applicable risks; (iii) federal income taxes on the future net revenues; and (iv) the likelihood of additional reserves, not considered proved at present, which may be recovered as a result of further exploration and development activities.

The following schedules present certain data pertaining to the Company's proved reserves.

Estimated Quantities of Proved Gas and Oil Reserves

	Oil (Bbls.)	Gas (MMcf.)
Proved Developed and Undeveloped Reserves:		
April 30, 1978	3,620,468	100,420
Revisions of previous estimates	(1,141,750)	(47,497)
Extensions, discoveries and other additions	277,827	27,929
Production	(258,864)	(5,479)
April 30, 1979	2,497,681	75,373
Revisions of previous estimates	(328,144)	(23,061)
Purchases of reserves-in-place	18,948	335
Extensions, discoveries and other additions	823,687	63,694
Production	(240,643)	(5,990)
April 30, 1980	2,771,529	110,351
Revisions of previous estimates	(689,951)	(32,239)
Extensions, discoveries and other additions	378,214	36,787
Production	(208,486)	(6,162)
April 30, 1981	2,251,306	108,737
Proved Developed Reserves:		
April 30, 1978	2,321,208	50,943
April 30, 1979	1,699,432	46,062
April 30, 1980	1,678,517	54,985
April 30, 1981	1,458,289	61,155

Estimated Future Net Revenues from Proved Gas and Oil Reserves

	Proved Developed and Undeveloped	Proved Developed
Future Net Revenues for Years Ending April 30:		
1982	$ 28,031,495	$ 36,960,652
1983	62,145,805	34,297,906
1984	64,044,259	29,622,854
Remainder	508,719,533	236,221,868
Total	$662,941,092	$337,103,280

Present Value, Discounted at 10%, of Estimated Future Net Revenues From Proved Gas and Oil Reserves

	Proved Developed and Undeveloped	Proved Developed
April 30, 1979	$ 73,251,000	$ 53,339,000
April 30, 1980	$242,555,000	$126,224,000
April 30, 1981	$368,237,000	$191,655,000

Changes in Present Value of Estimated Future Net Revenue from Proved Gas and Oil Reserves

Present value at April 30, 1979	$ 73,251,000
Additions and revisions, net of estimated future development and production costs	171,370,000
Purchase of reserves in place	573,000
Development costs incurred	11,381,000
Sales of gas and oil, net of related production costs	(14,020,000)
Present value at April 30, 1980	242,555,000
Additions and revisions, net of estimated future development and production costs	126,898,000
Sale of reserves	(68,000)
Development costs incurred	20,795,000
Sales of gas and oil, net of related production costs	(21,943,000)
Present value at April 30, 1981	$368,237,000

Summary of Gas and Oil Producing Activities on the Basis of Reserve Recognition Accounting

	Year Ended April 30	
	1981	1980
Additions and revisions to present value of estimated gas and oil reserves:		
Additions to estimated proved reserves, net of future development and production costs	$131,069,000	$136,852,000
Revisions to estimates of reserves proved in prior years:		
Changes in prices	79,152,000	48,710,000
Other	(107,265,000)	(21,482,000)
Accretion of discount	23,942,000	7,290,000
	126,898,000	171,370,000
Evaluated acquisition and exploration costs, including impairments	50,727,000	17,447,000
Additions and revisions to proved reserves over evaluated costs	76,171,000	153,923,000
Provision for income taxes	33,795,000	66,500,000
Results of gas and oil producing activities on the basis of reserve recognition accounting	$ 42,376,000	$ 87,423,000

Under Reserve Recognition Accounting (RRA), earnings are recognized upon discovery of proved reserves and upon subsequent revisions of values originally assigned to those discoveries. In contrast, earnings from gas and oil producing activities are not recognized in the Company's primary (historical cost) financial statements until the reserves are produced and sold. The Company's gas and oil producing operations, on a historical cost basis, produced a pre-tax contribution to profit of $16,002,000 for 1981 and $10,733,000 for 1980. Such results are included in the Company's primary financial statements. This compares to a pre-tax contribution of $76,171,000 for 1981 and $153,923,000 for 1980 on the basis of RRA.

The significant accounting policies which have been followed in preparing the Summary of Gas and Oil Producing Activities on the Basis of Reserve Recognition Accounting (RRA) are described in the following paragraphs.

Additions to estimated reserves represent future revenue, net of future development and production costs, from proved reserves added as a result of drilling exploratory and development wells.

Revisions to estimates of reserves proved in prior years include:

(a) The approximate effect of the changes in prices of gas and oil and production costs during the year.

(b) The net effect of all other changes affecting the RRA valuation not otherwise reported under revisions to estimates of reserves proved in prior years. These are composed primarily of the effects of changes in estimates and timing of future development costs, changes in timing of future production and revisions in estimated reserve quantities (primarily proved undeveloped reserves).

Accretion of discount was computed by applying the 10% discount factor to the present value of future net revenues. Accretion of discount gives recognition to the increase in the discounted reserve values resulting from the passage of time.

The costs of acquiring unproved properties and drilling exploratory wells are deferred until the properties are evaluated and determined to be either productive or non-productive, at which time they are charged to expense. At April 30, 1981 and 1980, accumulated exploration and acquisition costs of $7,192,000 and $8,015,000, respectively, were not fully evaluated and, therefore, have been deferred except for impairment provisions aggregating $1,565,000, all of which was provided during the fiscal year ended April 30, 1981.

Estimated future costs to develop proved reserves added during the year are expensed. As stated above, subsequent revisions to these cost estimates are included in revisions to estimates of reserves proved in prior years. Development costs incurred during the year that are associated with current additions to proved reserves are charged to expense.

The provision for income taxes includes income taxes currently payable and a deferred provision. The deferred provision is based on applying the year-end statutory income tax rates to the difference between (a) the RRA valuation of proved reserves and (b) the tax basis in proved gas and oil properties at year-end, and taking into account investment tax credits associated with future development costs and estimated statutory depletion associated with future production of proved reserves.

Report of Ernst & Whinney Independent Auditors

46

Board of Directors
Amarex, Inc.
Oklahoma City, Oklahoma

We have examined the consolidated balance sheets of Amarex, Inc. and subsidiaries as of April 30, 1981 and 1980, and the related consolidated statements of income, stockholders' equity, and changes in financial position for each of the three years in the period ended April 30, 1981. Our examinations were made in accordance with generally accepted auditing standards and, accordingly, included such tests of the accounting records and such other auditing procedures as we considered necessary in the circumstances.

In our opinion, the financial statements referred to above present fairly the consolidated financial position of Amarex, Inc. and subsidiaries at April 30, 1981 and 1980, and the consolidated results of their operations and changes in their financial position for each of the three years in the period ended April 30, 1981, in conformity with generally accepted accounting principles applied on a consistent basis.

Ernst & Whinney

Oklahoma City, Oklahoma
May 29, 1981

Appendix C

Example of Oil and Gas Lease

FORM 288-4-R (ORDER BY NUMBER)
Man.. Office Supply Co.--Printers--Okla. City B W

FORM 88 (PRODUCERS REVISED)

OIL AND GAS LEASE

(WITH POOLING CLAUSE)

THIS AGREEMENT, Entered into this the ________ day of ________, 19____

between ____

____ hereinafter called lessor,

and ____ hereinafter called lessee, does witness:

1. That lessor, for and in consideration of the sum of ________ Dollars in hand paid and of the covenants and agreements hereinafter contained to be performed by the lessee, has this day granted, leased, and let and by these presents does hereby grant, lease, and let exclusively unto the lessee the hereinafter described land, and with the right to unitize this lease or any part thereof with other oil and gas leases as to all or any part of the lands covered thereby as hereinafter provided, for the purpose of carrying on geological, geophysical and other exploratory work, including core drilling, and the drilling, mining, and operating for, producing, and saving all of the oil, gas, casinghead gas, casinghead gasoline and all other gases and their respective constituent vapors, and for constructing roads, laying pipe lines, building tanks, storing oil, building power stations, telephone lines and other structures thereon necessary or convenient for the economical operation of said land alone or conjointly with neighboring lands, to produce, save, take care of, and manufacture all of such substances, and

for housing and boarding employees, said tract of land with any reversionary rights therein being situated in the County of ________________

________________, State of ________________, and described as follows:

in Section ________, Township ________, Range ________, and containing ________ acres, more or less.

2. This lease shall remain in force for a term of ________ years and as long thereafter as oil, gas, casinghead gas, casinghead gasoline or any of the products covered by this lease is or can be produced.

3. The lessee shall deliver to lessor as royalty, free of cost, on the lease, or into the pipe line to which lessee may connect its wells the equal one-eighth part of all oil produced and saved from the leased premises, or at the lessee's option may pay to the lessor for such one-eighth royalty the market price for oil of like grade and gravity prevailing on the day such oil is run into the pipe line or into storage tanks.

4. The lessee shall pay to lessor for gas produced from any oil well and used by the lessee for the manufacture of gasoline or any other product as royalty 1/8 of the market value of such gas at the mouth of the well; if said gas is sold by the lessee, then as royalty 1/8 of the proceeds of the sale thereof at the mouth of the well. The lessee shall pay lessor as royalty 1/8 of the proceeds from the sale of gas as such at the mouth of the well where gas only is found and where such gas is not sold or used, lessee shall pay or tender annually at the end of each yearly period during which such gas is not sold or used, as royalty, an amount equal to the delay rental provided in paragraph 5 hereof, and while said royalty is so paid or tendered this lease shall be held as a producing lease under paragraph 2 hereof; the lessor to have gas free of charge from any gas well on the leased premises for stoves and inside lights in the principal dwelling house on said land by making his own connections with the well, the use of such gas to be at the lessor's sole risk and expense.

5. If operations for the drilling of a well for oil or gas are not commenced on said land on or before the ________ day of ________, 19____,

this lease shall terminate as to both parties, unless the lessee shall on or before said date pay or tender to the lessor or for the lessor's credit in the ________________

________________ Bank at ________________, or its successors, which Bank and its successors are the lessor's agent and shall continue as the depository of any and all sums payable under this lease regardless of changes of ownership in said land or in the oil and gas or in the rentals

to accrue hereunder, the sum of ________________ Dollars, which shall operate as a rental and cover the privilege of deferring the commencement of operations for drilling for a period of one year. In like manner and upon like payments or tenders the commencement of operations for drilling may further be deferred for like periods successively. All payments or tenders may be made by check or draft of lessee or any assignee thereof, mailed or delivered on or before the rental paying date, either direct to lessor or assigns or to said depository bank, and it is understood and agreed that the consideration first recited herein, the down payment, covers not only the privilege granted to the date when said first rental is payable as aforesaid, but also the lessee's option of extending that period as aforesaid and any and all other rights conferred. Notwithstanding the death of the lessor or his successors in interest, the payment or tender of rentals in the manner above shall be binding on the heirs, devisees, executors, and administrators of such persons.

6. If at any time prior to the discovery of oil or gas on this land and during the term of this lease, the lessee shall drill a dry hole, or holes on this land, this lease shall not terminate, provided operations for the drilling of a well shall be commenced by the next ensuing rental paying date, or provided the lessee begins or resumes the payment of rentals in the manner and amount hereinabove provided, and in this event the preceding paragraphs hereof governing the payment of rentals and the manner and effect thereof shall continue in force.

7. In case said lessor owns a less interest in the above described land than the entire and undivided fee simple estate therein then the royalties and rentals herein provided for shall be paid the said lessor only in the proportion which his interest bears to the whole and undivided fee. However, such rental shall be increased at the next succeeding rental anniversary after any reversion occurs to cover the interest so acquired.

8. The lessee shall have the right to use, free of cost, gas, oil and water found on said land for its operations thereon, except water from the wells of the lessor. When required by lessor, the lessee shall bury its pipe lines below plow depth and shall pay for damage caused by its operations to growing crops on said land. No well shall be drilled nearer than 200 feet to the house or barn now on said premises without written consent of the lessor. Lessee shall have the right at any time during, or after the expiration of, this lease to remove all machinery, fixtures, houses, buildings and other structures placed on said premises, including the right to draw and remove all casing, but lessee shall be under no obligation to do so, nor shall lessee be under any obligation to restore the surface to its original condition, where any alterations or changes were due to operations reasonably necessary under this lease.

9. If the estate of either party hereto is assigned (and the privilege of assigning in whole or in part is expressly allowed), the covenants hereof shall extend to the heirs, devisees, executors, administrators, successors, and assigns, but no change of ownership in the land or in the rentals or royalties or any sum due under this lease shall be binding on the lessee until it has been furnished with either the original recorded instrument of conveyance or a duly certified copy thereof or a certified copy of the will of any deceased owner and of the probate thereof, or certified copy of the proceedings showing appointment of an administrator for the estate of any deceased owner, whichever is appropriate, together with all original recorded instruments of conveyance or duly certified copies thereof necessary in showing a complete chain of title back of lessor to the full interest claimed, and all advance payments of rentals made hereunder before receipt of said documents shall be binding on any direct or indirect assignee, grantee, devisee, administrator, executor, or heir of lessor.

10. If the leased premises are now or shall hereafter be owned in severalty or in separate tracts, the premises nevertheless shall be developed and operated as one lease, and all royalties accruing hereunder shall be treated as an entirety and shall be divided among and paid to such separate owners in the proportion that the acreage owned by each separate owner bears to the entire leased acreage. There shall be no obligation on the part of the lessee to offset wells on separate tracts into which the land covered by this lease may be hereafter divided by sale, devise, descent or otherwise or to furnish separate measuring or receiving tanks. It is hereby agreed that in the event this lease shall be assigned as to a part or as to parts of the above described land and the holder or owner of any such part or parts shall make default in the payment of the proportionate part of the rent due from him or them, such default shall not operate to defeat or affect this lease insofar as it covers a part of said land upon which the lessee or any assignee hereof shall make due payment of said rentals.

11. Lessor hereby warrants and agrees to defend the title to the land herein described and agrees that the lessee, at its option, may pay and discharge in whole or in part any taxes, mortgages, or other liens existing, levied, or assessed on or against the above described lands and, in event it exercises such option, it shall be subrogated to the rights of any holder or holders thereof and may reimburse itself by applying to the discharge of any such mortgage, tax or other lien, any royalty or rentals accruing hereunder.

12. Notwithstanding anything in this lease contained to the contrary, it is expressly agreed that if lessee shall commence operations for drilling at any time while this lease is in force, this lease shall remain in force and its terms shall continue so long as such operations are prosecuted and, if production results therefrom, then as long as production continues.

13. If within the primary term of this lease, production on the leased premises shall cease from any cause, this lease shall not terminate provided operations for the drilling of a well shall be commenced before or on the next ensuing rental paying date; or provided lessee begins or resumes the payment of rentals in the manner and amount hereinbefore provided. If, after the expiration of the primary term of this lease, production on the leased premises shall cease from any cause, this lease shall not terminate provided lessee resumes operations for drilling a well within sixty (60) days from such cessation, and this lease shall remain in force during the prosecution of such operations and, if production results therefrom, then as long as production continues.

14. Lessee may at any time surrender or cancel this lease in whole or in part by delivering or mailing such release to the lessor, or by placing same of record in the proper county. In case said lease is surrendered and canceled as to only a portion of the acreage covered thereby, then all payments and liabilities thereafter accruing under the terms of said lease as to the portion canceled shall cease and determine and any rentals thereafter paid may be apportioned on an acreage basis, but as to the portion of the acreage not released the terms and provisions of this lease shall continue and remain in full force and effect for all purposes.

15. All provisions hereof, express or implied, shall be subject to all federal and state laws and the orders, rules, or regulations (and interpretations thereof) of all governmental agencies administering the same, and this lease shall not be in any way terminated wholly or partially nor shall the lessee be liable in damages for failure to comply with any of the express or implied provisions hereof if such failure accords with any such laws, orders, rules or regulations (or interpretations thereof). If lessee should be prevented during the last six months of the primary term hereof from drilling a well hereunder by the order of any constituted authority having jurisdiction thereover, or if lessee should be unable during said period to drill a well hereunder due to equipment necessary in the drilling thereof not being available on account of any cause, the primary term of this lease shall continue until six months after said order is suspended and/or said equipment is available, but the lessee shall pay delay rentals herein provided during such extended time.

16. Lessee, at its option, is hereby given the right and power to pool or combine the acreage covered by this lease or any portion thereof with other land, lease or leases in the immediate vicinity thereof, when in lessee's judgment it is necessary or advisable to do so in order to properly develop and operate said lease premises so as to promote the conservation of oil and gas or other minerals in and under and that may be produced from said premises, such pooling to be of tracts contiguous to one another and to be into a unit or units not exceeding 80 acres each in the event of an oil well, or into a unit or units not exceeding 640 acres each in the event of a gas well. Lessee shall execute in writing and record in the conveyance records of the county in which the land herein leased is situated an instrument identifying and describing the pooled acreage. The entire acreage so pooled into a tract or unit shall be treated, for all purposes except the payment of royalties on production from the pooled unit, as if it were included in this lease. If production is found on the pooled acreage, it shall be treated as if production is had from this lease, whether the well or wells be located on the premises covered by this lease or not. In lieu of the royalties elsewhere herein specified, lessor shall receive on production from a unit so pooled only such portion of the royalty stipulated herein as the amount of his acreage placed in the unit or his royalty interest therein on an acreage basis, bears to the total acreage so pooled in the particular unit involved.

17. This lease and all its terms, conditions and stipulations shall extend to and be binding on all successors of said lessor or lessee.

IN WITNESS WHEREOF, we sign the day and year first above written.

________________________ (SEAL) ________________________ (SEAL)

________________________ (SEAL) ________________________ (SEAL)

________________________ (SEAL) ________________________ (SEAL)

________________________ (SEAL) ________________________ (SEAL)

STATE OF____________________ }
COUNTY OF___________________ } SS

ACKNOWLEDGMENT. Applicable for lands in Oklahoma Kansas, Nebraska, North and South Dakota, Arizona, Colorado, Indiana, Mississippi, Oregon, Wyoming, and/or New Mexico.

BE IT REMEMBERED, That on this_____day of _______________ A. D., 19_____, before me, a Notary Public in and for said County and State, personally appeared____

____

__________________________to me known to be the identical person___ described in and who executed the within and foregoing instrument and acknowledged to me that_______executed the same as___________free and voluntary act and deed for the purposes therein set forth.

IN WITNESS WHEREOF, I have hereunto set my official signature and affixed my notarial seal, the day and year first above written.

My commission expires:______________________ ______________________________Notary Public

CORPORATION ACKNOWLEDGMENT **(Oklahoma Form)**

STATE OF______________________________County of__________________________, ss:

On this_________day of__________________________, A. D., 19_____, before me, the undersigned, a Notary Public in and for the county and state aforesaid, personally appeared_____________________________________ to me known to be the identical person who signed the name of the maker thereof to the within and foregoing instrument as its ___________President and acknowledged to me that__________executed the same as_____________free and voluntary act and deed, and as the free and voluntary act and deed of said corporation, for the uses and purposes therein set forth.

Given under my hand and seal the day and year last above written.

My commission expires:______________________ ______________________________Notary Public

When instrument is executed by a corporation, the corporate name must be shown and instrument signed by its President or Vice-President and attested by its Secretary or Assistant Secretary and the Corporate Seal affixed.

NOTARY ACKNOWLEDGMENT of SIGNATURE BY MARK **(Oklahoma Form)**

STATE OF______________________________County of__________________________, ss:

Before me, _________________________, a Notary Public in and for said County and State on this__________ day of___________________, 19_____, personally appeared______________________________________

____

to me known to be the identical person___ who executed the within and foregoing instrument by_______________mark in my presence and in the presence of____ as witnesses and acknowledged to me that____________executed the same as________________free and voluntary act and deed for the uses and purposes therein set forth.

In Witness Whereof, I have hereunto set my hand and official seal the day and year last above written.

My commission expires:______________________ ______________________________Notary Public

NOTE—The signature by mark of a lessor who cannot write his name must be witnessed by two witnesses, one of whom must write lessor's name.

FORM NO. 288-4-R
(ORDER BY NUMBER)

OIL AND GAS LEASE

FROM

TO

Dated___________________, 19___

Lot_____ Block_____ Addition

Township_______, Section_____ Range

County

No. of Acres__________ Terms

STATE OF____________ } ss.
____________County }

This instrument was filed for record on the

______day of_______________, 19___

at______o'clock_____M., and recorded

in Book__________of

at page________ Fee $_____

__________________County Clerk.

By___________________Deputy.

RETURN TO

TEXAS ACKNOWLEDGMENTS

THE STATE OF TEXAS, County of____, ss:
BEFORE ME, the undersigned, a Notary Public in and for said County and State, on this day personally appeared

____

known to me to be the person___ whose name_______subscribed to the foregoing instrument, and acknowledged to me that __he__ executed the same for the purposes and consideration therein expressed.

GIVEN UNDER MY HAND and the seal of this office, this______________day of___________________________, A.D., 19____

____

THE STATE OF TEXAS, County of____, ss:
BEFORE ME, the undersigned, a Notary Public in and for said County and State, on this day personally appeared

____wife of___ ____________________________________
known to me to be the person whose name is subscribed to the foregoing instrument, and having been examined by me privily and apart from her husband, and having the same fully explained to her, she, the said__________________________________ acknowledged such instrument to be her act and deed and declared that she had willingly signed the same for the purposes and consideration therein expressed, and that she did not wish to retract it.

GIVEN UNDER MY HAND and the seal of this office, this______________day of___________________________, A.D., 19___

____